Teaching Climate Science in the Elementary Classroom

Discover new ways to help elementary students engage with and understand the world around them through place-based, hope-filled learning about the causes, impacts, and responses to climate change. This book features foundational climate concepts, easily implementable activity plans, and inspiring examples of student engagement. Each chapter begins with a short vignette pulled from the author's considerable teaching experience in engaging students in concepts of climate change and climate justice, followed by content-focused sections and recommendations for student activities and projects. The author provides stories of hope-filled action to invite teachers to look for and reflect on similar narratives in their own communities. Sample units of study for grades K-5 show teachers how key ideas from each chapter come together into an instructional plan that incorporates the three dimensions of NGSS and can fit into the broader outline of their school year. This resource is an accessible tool to support any elementary educator in building their own knowledge base and integrating the important and timely issues of climate change into their classroom.

Stephanie Sisk-Hilton is a Professor of Elementary Education at San Francisco State University. Her research focuses on how children and teachers co-create science understanding and how issues of agency and belonging interact with learning. She has taught elementary and middle school in Prince George's County, MD, Atlanta, GA, Brooklyn, NY, and Oakland and Berkeley, CA. She also facilitated a backyard summer science camp for neighborhood children for ten years. She works extensively as a teacher professional developer, supporting elementary educators to develop ambitious, joyful, and hope-filled science learning experiences.

Teaching Climate Science in the Elementary Classroom

A Place-Based, Hope-Filled Approach to Understanding Earth's Systems

Stephanie Sisk-Hilton

Routledge
Taylor & Francis Group

NEW YORK AND LONDON

Designed cover image: cover artwork by Hannah Wheeler

First published 2024
by Routledge
605 Third Avenue, New York, NY 10158

and by Routledge
4 Park Square, Milton Park, Abingdon, Oxon, OX14 4RN

Routledge is an imprint of the Taylor & Francis Group, an informa business

ISBN: 978-1-032-49386-2 (hbk)
ISBN: 978-1-032-48492-1 (pbk)
ISBN: 978-1-003-39353-5 (ebk)

DOI: 10.4324/9781003393535

Typeset in Palatino
by codeMantra

To the students of the San Francisco State University Multiple Subjects Credential Program
Your commitment to education as a tool for liberation inspires this work and transforms our world every day.

Contents

Foreword

Dr. Kelley Lê

*Executive Director, UC-CSU Environmental and
Climate Change Literacy Projects (ECCLPs)
Author of Teaching Climate Change for Grades 6–12:
Empowering Science Teachers to Take on
the Climate Crisis Through NGSS (2021)*

I was driving around with my family on a gloomy morning in Los Angeles, when one of my kids began to ask me about things he noticed outside the car window. He asked, "Mom, didn't you say that there's bad air next to the freeway?" Looking at the 405 overpass, I responded back, "Yeah, why? What are you wondering about?" He then goes on to share about the houses, the people, and the preschool he sees near the freeway as large trucks and cars move loudly in the background. Among the many questions he posed, he made one definitive statement that stopped me in my tracks. He said, "That's not right. Everyone deserves good air." Despite what many teachers believe young children can handle, this came from my four-year-old son. He might not have known the chemical makeup of pollution, the secondary sources of emissions, or the devastating health impacts by name at this age, but he knew that bad air was bad for people in our community and everyone has a right to good air.

Although I was a high school classroom teacher and instructional coach for over a decade, I wrestled with how much to share in that moment as I witnessed my son witnessing injustice. So I did what I was professionally trained to do. I continued to ask him questions (with compassion and age awareness) to push his complex critical thinking skills, and I let him know that I didn't know the solution but that I was excited to look up how others found success to inspire both of us.

These days, I work directly with PK-12 educators who are looking for support to teach about climate change in ways that activate student agency. Dr. Sisk-Hilton's book is such a timely and important resource for elementary educators who are looking for tangible ways to integrate culturally relevant and solution-oriented ways of teaching climate change. As students continue to be disproportionately confronted with the impacts of this crisis, teachers can help to provide safe spaces to explore creative ways that position students as essential changemakers. We all have a critical role in solving climate change, and teaching about it is an essential climate action. We can engage our youth to help determine the path forward in ways that are just, equitable, and humane for all living things we share this planet with.

As Dr. Sisk-Hilton emphasizes a variety of research-based approaches for tackling climate change, it's important to remember that practices will only change so long as they are aligned to our personal beliefs and values. Every educator makes daily decisions in the classroom based on their own theories of action and change. These decisions are greatly influenced by one's own culture, identity, background, and schooling experiences among other important factors. When deciding what to include, omit, or adapt—reflect on your own teaching beliefs and values to gain an understanding of your current curriculum and pedagogical practices. Where you currently are in your teaching and where you aspire to go as you explore this book are great goal posts to continually revisit to track any changes over time.

Climate change is everywhere and all around us—to the point where many have become completely desensitized to news of worldwide tipping points, warning signs, and dire international reports from experts in the field. While some have the privilege of ignoring all of the above, our youth are more worried than ever about climate change and the current injustices further exacerbated by the crisis that impact their lived realities. When the general public rely on the media as their main source of information on climate change, there is bound to be confusion, content overload, too much "doom and gloom" messaging, infusion of politics, and more resulting in climate fatigue. This is why it is essential to provide a safe space for your students to explore, question, and discover culturally relevant solutions with someone they trust (that's you!). You might not know all the answers to climate change (and don't worry no one does), but that doesn't mean that you cannot begin to take action today to discover, learn with, and co-construct those learning experiences with students using research-based approaches and credible curricular resources.

What is widely shared by elementary educators is the high importance placed on core subjects such as math and literacy for students. In my experiences connecting with school leaders and students, I also commonly hear that students are most engaged when they have opportunities to explore science and engineering. Climate change is the perfect vehicle to teach students about core subjects because it can be the context in which those subjects are situated to make the content directly relevant to student's lives. The subject naturally develops their complex critical thinking skills because the solutions needed are open-ended (and approaches for one community might not work for another). Just as Dr. Sisk-Hilton shares in her book, we can center on hope, innovation, and potential green careers as we position students as the drivers of meaningful change when they engage as authentic scientists and engineers.

It's important to remember that you are not alone in your learning journey and picking up this book is an important first step. Find your community of practice whether it be at your school, district, community, or state to ensure that you feel well supported in taking on climate change through education for climate action. Reach out to local community-based organizations to support students as environmental stewards and position yourself as a partner in co-creating that learning space together. Although some teachers shy away from the human impacts of climate change, we encourage you to lean into sharing the scientific consensus with students. Yes, we are changing climate, *and* also yes, we can still change climate for the better to ensure a sustainable future for generations to come. It's up to all of us, and teachers make up the magnificent critical mass needed to curb climate change. Our students are looking to us for support, guidance, and safe spaces to learn about their critical role in curbing climate change.

We got this.

Acknowledgments

This work existed in my head, in conversation with colleagues, and in growing piles of lesson plans for a long time before it became a book. I am full of gratitude for the many people who have helped develop the ideas and encouraged me to share them. First, a big thanks to the Routledge team, especially Alex Andrews and Alexis O'Brien, who have supported and helped shape this project from start to finish.

My colleagues at San Francisco State University have contributed to my understanding through their work in climate justice education and equity-oriented pedagogy. I am grateful to my colleagues in the SF State Climate Justice Leadership Initiative and the Center for Science and Math Education. Thanks also to Daniel Meier for holding space for writing over many, many cups of coffee. Special thanks to Sarah Ferner and Jamie Chan, both of whom co-developed learning experiences featured in this book.

I am so grateful to the students I have had the honor to work with over the past 30 years. Thank you to the students and teachers who have allowed me to co-teach and co-learn how to build understanding of earth systems and climate justice in our local community. Much love and thanks to the (now young adult) children and families who were part of Camp Stephanie and enthusiastically engaged in place-based science during their summer vacations. Thank you also to the preservice and veteran teachers who have participated in developing and testing lessons that bring to life the principles of systems, place-based learning, and hope-filled action.

And thank you, as always, to my family. I was extraordinarily lucky to be born into a family that models living lives of service and hope-filled action. My parents, Michael and Patricia Sisk, are unfailingly supportive of all that I do. My sisters and fellow educators, Martha Wheeler and Joanna Sisk-Purvis, inspire me and remind me to laugh. And Maxwell, Juliet, and Phil: thank you for your love and enthusiasm, even when our home is overtaken with science supplies and Mom won't stop talking about the carbon cycle. I love and appreciate you more than I can express.

1

Do Little Kids Really Need to Be Thinking about Climate Change?

It's a Friday afternoon in January. For the past week, I've been working with a group of future teachers to develop understanding of how children learn about the natural world and how to support their interest and love for science. On this final afternoon, grade-level teams present their ideas for engaging elementary students in a science topic that is important, interesting, and not present in their school's curriculum. One group stands in front of their peers and explains why their third graders need to learn about climate change: the impacts are present in their lives already, they are getting lots of information through media and need help processing it, and they will be the adults who need to guide society through the next stage of this global crisis. Heads nod in agreement. Then, the group projects a slide, a photograph of a polar bear floating on a dislodged piece of ice in the arctic. They explain that they want to focus on how climate change is impacting animals since their students love animals so much, and they think this will help them care about the issue. I look around the room and

DOI: 10.4324/9781003393535-1

see a few students shifting in their seats, and a couple of them look away from the projected image.

The group shares their plan to first study polar bears as a whole class and then give student groups different animals to learn about and advocate for. They focus mostly on how the animals "fit" in their environment but also on how their habitat is changing as a result of our destabilizing climate. At the end of their presentation, the class gives some feedback: they agree that third graders will be excited to learn about animals, and they appreciate the social studies connections as children learn about the different parts of the world where animals live. Then, a classmate raises her hand and begins to speak tentatively. She says,

> I really like how this is about something that kids love, how they'll get to learn about different animals. But I'm a little worried that they might end up, you know, a little scared or even scarred because they can't really do anything to make it better. I mean, when I saw that picture of the polar bear, I got kind of sick in my stomach because the ice is melting, animals are starving, and what can I do sitting here in San Francisco? Riding my bike instead of driving to campus isn't going to keep that polar bear from dying.

The conversation that followed was the seed for this book. My pre-service teachers, and so many of us who work with elementary students, are eager to help children make meaning of our rapidly changing world. We know that climate change is one of the most critical issues facing the world right now and for the foreseeable future. We see the impacts that climate instability is having on the communities in which we and our children live: increasingly frequent and violent storms that bring destruction, years of drought that impact livelihoods of agricultural workers and global access to food, more intense wildfires that displace whole communities. We also see the fear and uncertainty in our students and feel it ourselves. We know that one of the most important roles of an elementary teacher is to help students feel safe so that they can build the confidence needed to take on challenges. Learning about climate change, while critically important, may not feel safe. Proposing that teachers share images of dying polar bears with third graders was concerning, but it's easy to see how this group of caring, novice teachers got to that point. Like their young students, they care about the living things in the world around them, and also like their students, they want to feel like they are doing something that makes a positive difference. So what might we do instead?

Do little children really need to be thinking about climate change at all? Perhaps we should just do our best to protect them from this knowledge for as long as possible. After all, in addition to being frightening for all of us, the processes that are resulting in global climate instability are complex even for adults. However, as my pre-service teachers noted, children are already aware of many impacts of climate change, through the lens of their everyday experiences. Children living in California often get more excited than adults when sudden rainstorms come, having spent their whole lives knowing drought and water restrictions. Children living near the Gulf of Mexico and along the Eastern Seaboard learn how to pack emergency kits and prepare for evacuation as more frequent hurricanes impact their hometowns. Our students hear about young people such as Greta Thunberg, who at a young age has reached a global audience with her advocacy for urgent and immediate climate action. Children are immersed in this changing world and exposed to the things that the adults around them value, think about, talk about, and worry about. They are exposed to the human experience of climate change, and they need help making meaning of it.

I have spent 30 years teaching science with elementary students and with adult learners. I still struggle every day with how to best teach for climate change understanding and justice in ways that support rather than scare, empower rather than disengage. For the past decade, I have had the incredible joy of working with pre-service and early career teachers to consider how to teach the building blocks of climate science and climate justice in ways that center children's safety and sense of power. I've also worked directly with children to try out and refine these ideas in classrooms and in a summer science camp that I ran for ten years. Over and over again, I am reminded that we are eager to learn when the learning helps us feel more competent, confident, and able to make a difference in our world. The primary aim of the book is to equip teachers with the scientific and socio-scientific knowledge needed to understand global climate change and local impacts while also providing concrete ideas for how to engage young children in these ideas through a lens of *hope-filled action*.

Guiding Principles

Three guiding principles will guide our exploration of teaching for climate science and climate justice in elementary school:

1   Coming to understand the world as a set of **interdependent systems** is the key to understanding the causes, impacts, and potential solutions to climate change.
2   **Place-based learning** grounds children's knowledge in the places, people, and systems that they are part of and allows them to take up ideas of climate justice based on their own experiences and in their communities.
3   Engaging in collective, **hope-filled action**, whether direct action in our communities or envisioning and describing future efforts, helps children develop strong identities as people who have the knowledge and power to enact positive change.

Each of these principles comes from what we know about how children learn and what supports children and adults to engage in difficult problems and work toward solutions. I'll discuss each principle briefly here, and they will come up in each chapter of the book as we explore specific topics related to climate change and climate justice.

FIGURE 1.1 Justice-based climate education framework.

Principle 1: Build Understanding of the World as Interdependent Systems

Understanding the world as a set of *interdependent systems* is the base of much scientific knowledge. And understanding how systems interact and impact one another is the key to understanding the causes and impacts of anthropogenic climate change as well as how potential solutions might work to reduce them. Each chapter of this book examines a system that is changing in ways that lead to climate instability, being impacted by climate change, or in most cases, both. While some of the details of these systems are beyond what we need to teach to young children, the core idea that everything on earth, including ourselves, is connected through systems that provide the resources needed to survive and thrive is both accessible and meaningful as children work to understand their place in the world.

Sometimes, science materials intended for young children focus on discrete facts (for example, names and basic traits of dinosaurs) or on a single phenomenon (for example, describing a rainy day) without positioning the information within a system. People used to believe that young children could only understand "concrete" ideas that they could see and touch, and so teachers were discouraged from trying to engage children in anything abstract. We now know that this isn't true, and that in fact even very young children are busy making connections and developing their understanding of the systems that underlie their world: how family members and neighbors are connected to one another; how a garden goes from planted seeds to vegetables ready to harvest thanks to soil, water, light, and human tending; how the funny symbols we call letters come together to form all the words in a favorite story. Not only *can* children understand systems, but it is what our brains do all the time.[1] When ideas are connected to one another, and especially when they are connected to things we already know and care about, our knowledge and understanding grow. When they are presented as discrete facts, unconnected to our lives and experiences, we struggle to remember them when the unit of study has concluded.[2]

Here's an example of going from fact-teaching to systems-based learning. I have a vague memory of learning about different components of soil in third or fourth grade. I think I remember it because we used magnifiers and got to look at actual soil instead of just reading about it in a book, something that did not happen too often in my elementary schooling. But I don't remember the point of the lesson or any of the details. When, as an adult in my 30s, I moved to a house with a backyard full of nearly impenetrable clay, I quickly (re-)learned the characteristics of sand, humus, and of course, clay, as I had a vested interest in improving the system that eventually became our backyard garden. Since my children grew up in that home and have helped with soil supplementation, planting, and weeding since they were little, they know far more about soil as part of a garden system than I did as a child.

When my son was in fourth grade, he came home and told us about doing the very lesson that I recalled from my own childhood. However, what he took from it was much more extensive than what I had. His teacher had embedded the soil lesson into a study of growing food, and they had examined soil components in order to decide what they should mix together to plant lettuce and other food crops. In fact, he was telling me about the lesson because he thought we needed to make some changes to our garden soil, adding more sand to improve the drainage. As a result of his teacher's contextualization of the lesson and his own life experience, he understood soil and its components as part of a *garden system* and so was easily able to connect this new information from school with what he already knew from his first-hand experience.

A Note about Terminology

Several terms will come up repeatedly throughout this book. Below are brief definitions in case some of these don't regularly come up in your everyday conversation!

Anthropogenic: This term means *caused by or originating in human activity*. The term "anthropogenic climate change" refers to the human-caused changes to global climate since the industrial revolution. Across the earth's full history, many factors have resulted in changes to global climate, but the urgent situation we are in now is almost entirely due to human activity, so that will be the focus of this book. When I use the term "climate change" in this book, I will always be referring specifically to anthropogenic climate change.

Greenhouse gases: We will discuss these more thoroughly in Chapter 3, but in brief, the term "greenhouse gas" refers to any molecule that retains or "traps" heat from the sun and holds it in the earth's atmosphere. Carbon dioxide, methane, and nitrous oxide are the greenhouse gases that human activity has directly increased in the atmosphere. Water vapor is the most prevalent greenhouse gas in our atmosphere, and human activity that directly increases the concentration of other greenhouse gases also increases water evaporation and thus the amount of water vapor in the air.

Climate Change/Climate Instability: There is some disagreement about the best term to refer to what is happening to earth's system as a result of the overabundance of greenhouse gases in the atmosphere and oceans due to human activity. While the earth is warming in terms of overall average temperature, the term "global warming" does not contain all of the impacts and can cause misconceptions. Broadly, we are experiencing increasingly unstable climate patterns as a result of anthropogenic changes to the atmosphere and ocean. I will use the terms "climate change" and "climate instability" more or less interchangeably to refer to the broad phenomenon, knowing that these too are imperfect terms, but the ones currently in widest use.

Climate Justice: As with so many global problems, the people who are most vulnerable to the impacts of climate change are those who already experience poverty, food insecurity, sub-standard housing, and neighborhoods that have been exposed to toxic chemicals from industry and agriculture. Approaching the issues of climate change from a justice perspective means prioritizing immediate and long-term well-being for already vulnerable communities so that we move toward a sustainable and fair world for everyone rather than protection from climate change impacts being something only available to a few. Throughout the book, I emphasize community rather than individual action in part because justice-based responses to climate change require that we move beyond notions of individual action that are generally centered around affluence toward community-based frameworks that work toward systemic change and well-being for all.

Principle 2: Ground Learning in Place-Based Education

A second component that will help us teach for climate change and climate justice in developmentally supportive ways is grounding our work in *place-based education*. This means designing learning experiences that center children's communities and lived experiences. Teaching about climate change is challenging because of the magnitude of the problem and how powerless to act we often feel. But there is work being done in every community to build resiliency, reduce greenhouse gas emissions, and improve well-being of people and ecosystems. When children know a lot about the world around them and are supported to engage in their communities, they feel powerful and able to make change. Children who live near a stream have a vested interest in keeping it free of plastics and other waste. Children in communities impacted by drought are eager to develop and test water catching and storage devices. Concepts that might feel "too big" for young children when approached out of context become meaningful and accessible when tied to children's communities and experiences.

Children are able to clearly understand issues of climate justice as they impact their community, and in a local context, they have power to help make or advocate for change.[3] Children who see that some parts of their city are rich with trees while other areas are primarily concrete can measure differences in temperature and actually play a role in "tree justice" through helping to plant them or by writing letters to city leaders to share their findings. Children who have experienced the impact of big storms can find a feeling of safety in learning about how we can protect communities from storm surges and in designing model, storm-resilient neighborhoods. Helping children understand and take action in their own place helps to connect them to broader ecological and societal systems.

Place-based learning is a starting point, not a boundary, in coming to understand the broader world. Children should absolutely learn about places and people far different from the ones they know, from indigenous communities working to preserve their tropical rainforest homes to deep-ocean ecosystems that contain organisms that seem to come from science fiction. But the places children know well and experience every day ground them in connection to earth's systems and allow them to enact care on a child-friendly scale. And knowing their own place deeply allows children to make connections to faraway places and to people who are also working to make their own communities climate-resilient.

Principle 3: Support Children to Engage in Collective, Hope-Filled Action

Finally, one of the most effective ways to teach for empowerment rather than fear is to help children find ways to engage in *hope-filled, community-focused action*. This means developing children's sense of agency by supporting them to use their knowledge to make positive change. You'll notice throughout this book that I de-prioritize solitary, individual action and focus instead on community-based responses. This is for two reasons that bring together issues of power, impact, and justice. First, individual change by folks who have the privilege and resources to make changes, say to their diet or mode of transit, will not by itself result in change at the scale needed to mitigate the impacts of climate change. I can choose to adopt a more sustainable diet for myself, but one person's dietary choices does little to impact global greenhouse gas emissions or sustainability of our food system. On the other hand, helping to develop or advocate for community gardens and farmers' markets that provide access to fresh, local produce for folks experiencing food insecurity can grow into an effort that noticeably impacts both greenhouse gas emissions and community well-being. That's not to say that individual actions don't matter. I do, in fact, try to make sustainable choices in terms of diet, household energy use, and modes of transportation, and this helps to keep me engaged in climate action. But community-focused actions help us better consider social and ecological systems, and systems changes have more potential to be both just and transformative.

Second, I suggest that hope-filled actions be rooted in community solutions because children have limited power to enact individual change. For the most part, children aren't the ones deciding what or how much food is purchased in their household, whether they bike, walk, drive, or take a bus to school, whether their home's energy comes from solar panels or oil, or a host of other actions that we associate with being "climate friendly." And their grown-ups may not have abundant choices in these areas either. If you live in a community where schools are far from children's homes and the weather is harsh, commuting by car may be the only reasonable "choice." If a family is struggling to have enough money for rent and food, they aren't likely to purchase an electric car, nor would that be an advisable use of limited resources. If food budgets are limited, dietary decisions will be based on affordability, not sustainability. Suggesting that each of us are individually responsible for making the best possible climate-friendly decisions when we must consider so many other factors is unrealistic, and for children, it can cause shame and feelings of powerlessness. But children *can* help plant or tend to a street tree as part of a citywide effort to reduce the heat island effect. They *can* design model houses that keep people safe from storm surges and learn about people working to build more resilient coastal communities. When we zoom out from the individual to the community, we increase both impact and resilience by connecting our individual efforts and focusing on what can be done by working together.

Prioritizing Children's Feelings of Safety in a Scary World

Related to the idea of hope-filled action is a boundary we need to place on what content is introduced at the elementary school level. Children can and should learn big ideas, but we also need to protect them from scary ideas they cannot yet do anything about. When the human brain encounters frightening ideas, especially ones that may threaten our own or our loved one's safety, it shuts down. Scientists sometimes call this the "amygdala hijack." The amygdala is the oldest part of the human brain, and its responsibility is to keep us alive. So when danger pops up, the amygdala essentially overrules the problem-solving, innovating, understanding parts of our brain and tells us to flee to safety (or to fight, thus the well-known term "flight or fight"). Too often, climate news, much like the stranded polar bear image in the opening story, terrifies us and makes us feel powerless, and so we "flee" by disengaging.[4] Effective education for climate change and climate justice must come from a place of hope so that our children, and we ourselves, are able to remain committed to both learning and action.

Adults who fully understand the causes and impacts of human-caused climate change can maintain feelings of efficacy by engaging in and advocating for local and global changes needed to bring earth's systems back into balance. But when children learn about upsetting things that they cannot (yet) play a role in changing, they are understandably frozen with fear. It is our job as educators to have a depth of understanding of both climate content *and* child development, so that we can make informed choices about how to engage children in learning that will build understanding of climate change while also protecting them from information that they don't yet have the knowledge base to process through a hope-filled action framework.

You will notice that the science I cover in this book is aimed at adults, and the classroom connections don't always directly take up all aspects of that science. This is intentional, since the goal is to both increase adult knowledge and provide child-friendly entries into the content. So I will not be suggesting a first-grade study of melting arctic ice sheets, but I will recommend telling children stories of people who are using kelp farming to pull carbon from the atmosphere. And I will encourage you to find stories of justice-based climate action in your own communities so that children can be inspired and learn from leaders who are addressing issues that they can actually see and potentially be a part of. By focusing on understanding systems, building an empowered sense of place, and engaging in hope-filled action, we can reduce fear and set the stage for lifelong learning about our connected earth and how to better support our planet and its inhabitants.

Structure of This Book

Each of the chapters in this book contains several elements that ask you to move between the role of learner and educational designer. The chapters begin with a story of practice from my own teaching, to provide a glimpse into how children and teachers might take up issues of climate change and climate justice at a variety of grade levels. I share these stories *not* to provide a set of "perfect" examples! In fact, as I reviewed my years of teacher journals, student work samples, and lesson plans to create these vignettes, I found myself thinking of all the ways I could have taught the units more effectively and what I will do next time. I've invited you into my practice as a starting point, and I hope you will build your own set of stories of children engaged in hope-filled learning about global systems and our roles in them.

The body of most chapters moves back and forth between brief scientific explanations of core concepts related to the causes, impacts, and potential solutions to climate change and sections called "classroom connections" that suggest ways to use this adult-level knowledge, sometimes directly and often indirectly, to build children's understanding. The classroom connections serve as examples of how, by more fully understanding the causes, impacts, and responses to climate change ourselves, educators can design "building block" experiences for young children that will allow them to construct understanding over time. I also highlight stories of scientists and activists, including children and youth, who are engaged in improving their communities through climate action that is accessible to young children.

Most of the classroom connection sections are intentionally not grade-level specific for two reasons. First, while schools usually divide children by age, this is not really an indication of how learning happens. All of the classroom connections are based on learning experiences I have facilitated with children, and most of them have been implemented in both lower (kindergarten-grade 2) and upper (grades 3–6) elementary classrooms or in mixed-age environments. Of course, the level of detail and teacher supports will often look quite different in kindergarten versus a fifth-grade class. But teachers are experts at adjusting learning activities to meet the needs of their specific students, and age is rarely the most important factor. Second, I've tried to write these sections in a way that is open-ended enough that you can adjust them to the specific curricular requirements of your setting. They are meant to be idea generators, not recipes, and I encourage you to work with colleagues to bring some of them to life in ways that make sense for your students in your community.

Finally, most chapters include an example unit outline showing how the suggested content and classroom connections might fit within the Next Generation Science Standards (NGSS), the science education framework used in a majority of US states as of this writing. Depending on your particular context, you may be looking for ways to supplement your district-adopted science curriculum, or you may be essentially starting from scratch. The NGSS-based summary is meant to serve as a starting point as you and your school team develop a progression toward climate literacy in alignment with science learning goals across the elementary years.

I believe in the power of teachers to be agents of transformation in their classrooms and in our broader school and community systems. I am not writing a "how to" book but rather sharing a set of ideas that may help you as you consider how to engage in transformational teaching in your own community. I encourage you to seek out colleagues who are also committed to this work. My own practice has developed almost entirely in collaboration with other educators ranging from grade-level teams to university-school partnerships to statewide and national initiatives that introduced me to educators with similar goals and inspiring energy that fuels my own. Teaching, like all climate justice work, is most sustainable and impactful when done in community, and I hope that this book will help your community of educators.

A Note about Stories of Practice

Finally, I want to note that the stories of my own teaching practice throughout this book are based on copious documentation, but I have taken liberties with putting things in quotation marks, to aid the narrative structure. Because my academic work uses the teacher-researcher model, most of my teaching is heavily documented through lesson plans (yes, I still write them), reflective journaling, samples of student work, and often notes about class conversations written quickly in real time. But in most cases, class sessions were not audio or video-recorded, so the dialogues in some of the stories are re-created based on these sources. I am deeply grateful to my students, from kindergarteners through doctoral candidates, who willingly, often joyfully, engage in learning with me, and I hope that the way I've approached telling these stories helps bring their brilliance, excitement, and commitment to hope-filled action to life.

Notes

1 Gelman, S. A. (2009). Learning from others: Children's construction of concepts. *Annual review of psychology, 60,* 115–140.
2 National Research Council. (2007). *Taking science to school: Learning and teaching science in grades K-8.* National Academies Press.
3 Davis, N. R., & Schaeffer, J. (2019). Troubling troubled waters in elementary science education: Politics, ethics & black children's conceptions of water [justice] in the era of flint. *Cognition and Instruction, 37*(3), 367–389.
4 Fartachuk, Y., Wolf, M., Nghe, A., Brown, K., Becker, C., & Deckers, C. (2022). Mother earth and her anxious children. *Gray Matters Temple University,* Spring 2022. https://greymattersjournaltu.org/issue-3/mother-earth-and-her-anxious-children

2

Interconnectedness

Grown Up Science	Hope-Filled Classroom Connections
• What does it mean to think of Earth in terms of systems?	• Pollinator gardens as systems of living and non-living things
• How are the Earth's "spheres" of living and non-living elements interconnected?	• Considering "where am I in the story?" to find one's place in natural systems
• How do human brains organize information as systems using schema and story?	• Stories of systems and of hope-filled action to support children's schema as scientists and changemakers
• How is a systems orientation to understanding climate science connected to issues of climate justice?	

"Stephanie, Stephanie, come look! There's bees!" Two first graders have popped into the classroom during recess, as I set up for our upcoming science lesson. I step outside, and they pull me over to a plant that has recently started to bloom. Sure enough, there are several bees buzzing around it. One of the children correctly identifies the plant as lavender and wonders if the bees can smell it. Another says, "can we tell everyone we need to plant lavender

DOI: 10.4324/9781003393535-2

so we help the bees?" I tell them they can share their ideas at the beginning of science time, and they decide to come inside and make a "save the bees" sign to help them present.

This class's fascination with pollinators has developed over the past two months as the children have become keen observers of natural life in their mostly paved schoolyard, where a few landscape plants and a small school garden provide glimpses into the connection between plants and animals, especially insects. Earlier in the year, their primary teacher told me she planned to bring in caterpillars so they could observe their transformation into butterflies. I taught supplemental science lessons just once a week, so we decided I would focus on helping them understand the role of pollinators in ecosystems and take action to create a pollinator-friendly habitat near our school.

To begin our study, we took an insect walk in the school-yard, and even in this less than ideal habitat, students noticed plenty of activity: a line of ants headed to a forgotten piece of food, small flies near the trash can, and creatures they identified as insects (actually isopods or "roly polies") in the wood chips under the play structure. It was late winter, so they didn't see any evidence of insect pollinators, but when we brainstormed more insects they knew about, children mentioned both bees and butterflies. They eagerly told me what they were learning about the caterpillars that had recently arrived in their classroom. There was a small garden on campus where older students were growing lettuce, beans, and other food crops. We invited fourth graders to visit the first graders, and they told them how insects were needed to pollinate the flowers of the beans, or there would be no beans at all. While this part of the process was still a couple of months away, some children began checking the plants for evidence of pollinators, or of beans, every day at recess.

While the lead teacher taught them about butterflies, I focused on bees and their role in pollination. We used chenille stems to make model bee bodies and baking powder on paper flowers to model pollen, and children acted out the process of bees moving from flower to flower, spreading pollen as they collected their food. I went to the public library, where a children's librarian helped me gather enough books about bees, butterflies, and gardens that every child could select one to read, alone or with a partner. They illustrated their findings, and after they shared with the class, we posted these around the room.

One of the after-school instructors also worked in community gardens, so he visited the classroom during science time to explain that there were fewer and fewer bees each year, but we need bees in our gardens and farms and in natural habitats, as they are such an important part of the life cycles of many plants. He also showed them pictures of different types of bees and explained that each of them were adapted to certain types of plants.

One afternoon, we took the children to a small patch of dirt between the school gate and the sidewalk that needed care (and that we had permission to transform). We asked if there was a way to turn this into a place where bees and butterflies could get what they needed to live. Students drew plans of what they thought the area should look like. A parent who worked with a local plant nursery collected their ideas and tried to get plants similar to what the children requested. As student pairs spent a couple of our weekly sessions learning about the native plants that the parent volunteer had obtained, the sun came out, and the schoolyard plants began to bloom. They eagerly reported each new citing of a bee or butterfly.

Before they planted their native plant garden, the children made laminated signs to explain to visitors what each plant was and which insects used it for food and pollinated it. Because the garden was just outside the school grounds, they were able to see neighbors pause and read the signs and look at the plants. They had taken a small, unused spot of earth and transformed it into a life-giving space not only for plants and insects but also for the people in their community.

Systems Connect Everything on Earth

Have you ever sat quietly for a long period in a natural place such as a forest, a meadow, or a beach? At first, you may see the most prominent features of the landscape: tall trees filtering the sunlight or a flat field of grass or crashing waves. Sit longer, and smaller, more subtle things come into focus: the sound of birds hidden from view, insects crawling on the ground or buzzing around flowers, the differing shapes and shadings of leaves. As you breathe and take in the natural world around you, you may feel a deep sense of connection to this place. I feel this when I spend time in the deciduous forests of my youth but also in places that introduce me to new environments and natural systems, such as the Costa Rican rainforest or the desert regions of California.

One of the most important concepts in understanding the transformation that earth is undergoing right now is that everything on earth is, in some way, connected to everything else. The earth is a system of interrelated parts and processes, and within this huge, global system, there are countless smaller systems that sustain life and planetary stability. Water that evaporates from the surface of the ocean into the sky can affect the weather patterns on inland areas where few people have ever seen the ocean. Snow that falls in the mountains hundreds of miles from my home melts in spring and summer and provides most of the water for my region. A late season freeze in Florida impacts communities in Iowa as they are less able to access citrus fruit. Deforestation in the rainforest regions of South America impact the amount of global carbon in the atmosphere. When a community works to eliminate plastic waste from entering waterways, this can create cleaner water far out in the ocean or hundreds of miles downstream.

Scientists often refer to the global system in terms of five "spheres," which refer to types of materials that contribute to earth's life-sustaining systems. The **biosphere** refers to all living organisms on earth. The **geosphere** is an overarching category that refers to all of the non-living components of earths' systems. Within the geosphere are four sub-categories. The **lithosphere** refers to the rocks and soil that form the earth's surface and interior. The **hydrosphere** is the network of water that is present across the earth, from the ocean to rivers and lakes. The **atmosphere** refers to layers of gases that surround the earth's surface, providing temperature regulation, water circulation, and access to gases needed by living things. The **cryosphere** is the ice that

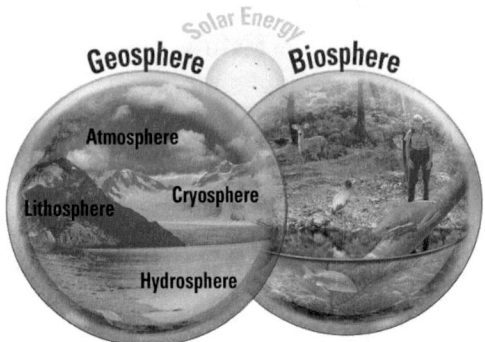

FIGURE 2.1 The earth's systems as five spheres. Image courtesy of USGS, Graphic Design by James A. Tomberlin. Williams, R. S. (2012). The earth's dynamic cryosphere and the earth system. United States Geological Survey. https://pubs.usgs.gov/pp/p1386a/pdf/notes/1-8hydrocycle_508.pdf

covers the polar regions. Because ice is a form of water, some resources do not include this as a separate "sphere," but the role of ice in global climate is distinct enough from that of liquid water and water vapor to warrant separate consideration.

With the exception of ice, all of these components are present in some form in every ecosystem on the planet, and their interactions support life on earth. As I type this, I am looking out my window at rain that is falling from the atmosphere, landing on rocks and soil as well as on the built environment of houses, sidewalk, and street. Some of this water will make its way underground to add to groundwater stores, and some will flow into the San Francisco Bay and the Pacific Ocean. Plants will use the water to support their growth, and animals will drink from water sources that are replenished by the rain. Later in the year, melting snow will make its way from the Sierra Mountains through rivers toward my part of the world, filling reservoirs and providing water for my community to grow our food, sustain our bodies, clean things, and support industrial processes.

The rain falling outside my window hasn't always been in the form of water vapor directly above my house! At some point, it may have been part of the ocean water. It may have been released by plants as they engage in respiration to use stored energy. The elements that make up the water molecules that I'm seeing now may have made their way around the world over millions of years of cycling through different parts of the earth's systems. There is a wonderful children's book, *One Well*, that helps children imagine how water in their environment may have circulated through different places and times all over the earth.[1]

In every place on earth, the five spheres interact to create environmental conditions in which life that is adapted to those conditions can thrive. Some places on earth are teeming with life because of the environmental conditions. For instance, rainforest regions have abundant water, warm atmospheric conditions year-round, and a topography that allows for an enormous array of plant and animal life to develop and survive. Other areas are hospitable to only a few specially adapted organisms. Deserts, with their scarce water, have limited plant life, and since fewer nutrients from decayed organisms enter the soil, this further reduces plant growth. Less water vapor in the air results in temperature extremes, with very hot temperatures during the day giving way to often extreme cold when the sun is not shining. The lack of trees and other forms of shade means that only plants and animals that have adaptations to protect from temperature extremes and rapid water evaporation are able to survive.

Whether crowded with living things or providing conditions for only a few hardy organisms, when any environment experiences significant change to the characteristics of one of the spheres, this impacts what is able to live and thrive in that area. We'll discuss specific changes in each of the chapters of this book, and we'll also explore how communities can work to reduce or reverse the human causes and impacts of disruption to earth's systems.

Humans Are a Part of, Not Apart from, Earth's Systems

We sometimes think of humans as residing outside these global systems because we are able to use technology and innovation to survive in conditions that would not seem conducive to living things, such as cities where the geosphere and hydrosphere are often no longer very visible. But despite our ability to create and wear layers of clothing, build shelter from weather, cultivate food crops that aren't native to an area, and transport water to arid lands, we are still inexplicably tied to natural systems on a local and global scale.

Let's look at an example of humans' interactions with the hydrosphere (and others, as all of the spheres are fundamentally connected). In the Western region of the United States, there are constant political battles over access to water, including water from rivers such as the Colorado as well as groundwater that supports agriculture. As population, agriculture, and industry have grown, water use has stretched beyond what the natural system can sustain. With climate instability increasing periods of drought and flood, we need to develop new ways of being a part of this hydrologic system.

Adopting lower water use practices is certainly a start, and so too are innovations to adjust to the unpredictable levels of precipitation. Humans can use engineering and geological knowledge to better capture stormwater to replenish the groundwater in underground aquifers, rather than have the majority of it head to the oceans in the form of runoff, which often brings with it pollutants and sediments that harm aquatic environments. Such strategies must take into account their impact not just on a single community but on an entire watershed, as changing how water flows in one place impacts how much water ends up downstream as well, impacting habitats for plants and animals as well as human communities. In describing how to best develop new approaches to human water use, Sharon Megdal, a water resources researcher, explains, "You have to look at this in the context of your whole system: Your hydrologic system, your legal system, and then how the users are using the water."[2]

Smaller scale examples of our connection to natural systems abound, and these are a starting point for connecting global systems to students' everyday experiences. Several years ago, my students noticed that there were more and more stories of coyotes coming into local neighborhoods, even in cities, sometimes eating household pets. Several students expressed fear of being attacked by coyotes, so I decided we should learn more about them in order to address fear of the unknown with increased understanding.

In their research, students learned that scientists have several ideas about the proliferation of coyotes in cities. Habitat destruction likely plays a role, with coyotes having less access to unpopulated areas to hunt, but there had not been a huge change in that during my students' elementary years. Our region was several years into the worse drought in the area's known history, almost certainly exacerbated by human-caused climate instability. The drought was leading to less water and therefore fewer animals, in natural areas, and it was also a factor in larger and longer-burning wildfires. This was more likely the most immediate explanation.[3] When coyotes ventured into human neighborhoods, they might find water in the form of pools, fountains, dripping hoses, and other things created to support human communities. And while household pets were understandably my students' primary concern, human communities also host rodents and other animals that we consider pests, but which coyotes, who are not picky eaters, might consider dinner.

As my students gained knowledge of how coyotes are connected to the same natural systems as we are, their fear mitigated somewhat. They absolutely did need to stay far from any wild animal for their safety and the animal's, and those with cats talked with their parents about keeping them indoors (also important for maintaining songbird populations). But understanding what might be bringing coyotes into our neighborhoods also helped them develop empathy for these creatures, and they wanted to learn more about protecting their habitats. A naturalist from our regional park system was already scheduled to do a program with them, and when he heard about their coyote research, he took time to answer questions about fire risk mitigation and habitat restoration. His hope-filled, action-based response clearly impacted my students. Even though they could not immediately change this problem, in learning about this part of our local ecosystem, they developed understanding of connections between living and non-living things, and they saw that solutions must address not just our immediate needs but also those of other organisms and the resources we all depend on.

Classroom Connections: How Am I Connected to the World around Me?

Learning about a local plant or animal in the context of local ecosystems helps children understand that we are all part of interconnected systems. In the opening story, the first-grade teacher and I expanded a unit on butterfly life cycles to help children understand the role of pollinators in our local area. They used their growing knowledge to improve a small area in our community in hopes of increasing local pollinators and supporting human well-being by adding nature to our urban area.

We often teach about living things in ways that de-emphasizes both biodiversity and connection to systems. For instance, children might learn basic butterfly life cycle stages and that they eat flower nectar (in their adult phase). But there are more than 17,000 species of butterflies in the world![4] Each of these has co-evolved with the plants in their habitat that serve as food at different life stages. Place-based education challenges generic "all about butterflies" approaches by grounding learning in the specifics of a local community and environment. These stories and experiences rooted in place help children see connections between themselves and the broader world.

The California pipevine swallowtail, a butterfly species local to the place I live, feeds on only one plant, the California pipevine, in its caterpillar phase. Both have become rare in urban areas due to habitat loss. A local biologist, Tim Wong, has devoted years to restoring this part of the ecological system in the city of San Francisco. He collected pipevine clippings from the San Francisco Botanical Garden (with permission) and used these to grow new plants in his backyard. He reached out to community members to find caterpillars in areas where swallowtails are more common. He created a protective enclosure to protect the caterpillars and increase mating opportunities for adult swallowtails. His multi-year effort has re-introduced thousands of these insects to the city, along with the native habitat in which they thrive.[5]

In recent work I've done in collaboration with my colleague Jamie Chan, we have introduced children to the special relationship between the California pipevine swallowtail and the pipevine plant, moving away from learning about concepts like pollination, life cycle, and plant/animal relations *in general* toward understanding them in the context of this local, tangible system. In the process, children learn about a local scientist and community member whose work brings the impact of science and activism to life in accessible ways.

Re-focusing learning toward local systems also facilitates hope-filled action rather than just "learning about." In the example above, once children grow native plants that attract and depend on local pollinators, opportunities abound to observe natural cycles and systems. What parts of your curriculum could be modified to focus on a local habitat or natural system? And how might you invite children to consider their role within this system? We'll take that up next in considering the role of story in building understanding.

Our Brains Organize Knowledge as an Interconnected System: Schema and Stories

Human brains can process and store vast amounts of knowledge. By the time most children enter school, they are fluent in at least one spoken language, able to express both concrete knowledge such as how to get ready for school and abstract ideas such as what love is. Many five- and six-year olds have already begun to learn not just one or more spoken languages, but also the written systems used to record and transmit information. They are practicing a whole host of skills from eating with a fork or chopsticks to comforting a sad friend or sibling.

How do our brains manage to remember so much? And why am I able to remember almost word for word stories my parents told me as a child, but I seem to have entirely forgotten the content of many of my college classes? Our brains are, themselves, complex systems designed to store and retrieve information that we need in the context in which we live. When new information is encoded in our brain, we search for connection to what we already know and understand, regardless of whether our current knowledge is accurate. As we connect related knowledge and experiences, we form networks known as "schema." We are far more likely to remember and be able to retrieve knowledge connected to rich schema than isolated facts.[6] I have spent my adult life studying children's science learning, so my schema in that area are large and well developed. When I hear something new regarding science education, I quickly connect it to and compare it to my current understanding. On the other hand, I have very little knowledge of sports teams, and when I hear friends discussing statistics of games from weeks or even years before, I'm shocked that they can recall this information. My professional sports schema are small and undeveloped, so I have less to draw upon when faced with new information in that area.

Knowing how our brains process information makes clear that teaching for systems understanding not only helps children see the earth and everything on it as a set of interrelated systems but also increases their ability to retain what they learn at a young age and connect it to new and growing understanding across time. In Chapter 1, I contrasted my own vague memory of learning about soil types in elementary school to my son's experience learning about soil in the context of a garden system and using that knowledge to grow food with his class. His teacher connected "knowing about" soil components to understanding a system that students directly experienced, observed, and modified. Because of the poor soil in our backyard garden, my son was able to further connect the garden schema he was developing in school to knowledge from weekend chores and time spent playing outdoors. His knowledge of soil as a fourth grader was far more resilient than my own at the same age, and I credit his fourth-grade teacher's focus on understanding natural systems with much of his ongoing interest as a young adult in addressing climate change through re-envisioning land use.

Story is another critical part of how our brains organize and retain information, and this connects to the idea of schema. Human brains seem to be "wired" for story, meaning that we remember stories much better than we remember isolated pieces of information. For thousands of years, before written literacy was widespread, and still today in families and communities across the world, elders transmit knowledge to future generations and build understanding of the community's values and place in the world through stories.[7]

Those of us who work with children don't need to be told that stories are powerful. We see the way children light up when it's time to listen to a story. Small bodies edge closer, eager to hear what comes next. But sometimes, we forget that story and science are interconnected. There is a story to how water travels in the atmosphere, falls as snow on tall mountains, melts as the sun warms the land in Springtime, and flows through rivers toward communities that then use it for drinking water. There are countless stories contained in a forest, from the story of the life of a single tree over hundreds of years to the incredible stories of how plants and mycelium form networks that aid each other's growth and survival. When we look for the stories in climate-aware science, we find systems. We also find characters, whether trees or water droplets or humans or earthworms, and our brains are also wired to care about the characters in a story. Moving toward more narrative approaches to engaging with the natural world helps children feel more connected to the science and more likely to remember.

Classroom Connections: Where Am I in the Story?

For some of the novice teachers I work with, the thought that science is a system of stories is a revelation. When I challenge them to use story as a primary teaching tool in a science lesson, they come to our seminar with their eyes lit up, sharing how enthusiastic their students were when they approached a challenging science concept as a story of people in connection to the natural world.

Several of the units I highlight in this book build on the power of story to engage children in learning about systems. For instance, in the opening vignette in Chapter 3, children act out the story of the carbon cycle and human's impact on it. When I first taught that unit, this was the only way in which I used story to support learning. But as I tried to encourage my students to imagine bold responses to re-balance the carbon cycle, I found that stories of people engaged in community-based action were equally powerful in supporting systems understanding. For instance, we watched a video about a construction company that was switching to carbon-absorbing concrete. The story of the people who developed the concrete was exciting, and so too was imagining how the construction workers installing this might feel knowing that their work was improving not just their community but the carbon balance in the world. One child wondered if people could tell that carbon was being absorbed by the concrete and suggested developing a version that changed color, like a mood ring, to help people see the invisible good the concrete was doing. In imagining this innovation in the context of people and community, they had to draw upon their knowledge of the carbon cycle system as well as consider how people, as part of social and economic systems, were engaged with and impacted by this innovation.

One of my favorite prompts for children is "where am I in the story?" I first learned this prompt as part of an instructional approach used to share stories from my religious tradition, but I have found it to be equally powerful in helping children to consider how they "fit" in natural systems. Returning to the story of bees as pollinators, consider how children might take up the idea of "where am I in the story?" They might identify ways they are similar to one of the "characters." One child might say she is like the bee because she likes to run around in gardens, and another might say they are like the sun because their parent says they give energy to others. On a less metaphorical level, children might think about their personal role in the story, perhaps in providing and advocating for bee-friendly habitats. Asking children to consider where they are in any story encourages them to consider their role in a system.

When you are planning to teach a complex concept with your students, try looking for stories as entry points and throughlines. An example for younger grade children (although this is compelling to older students as well) is the story of *The Soda Bottle School*, a movement in Guatemala to re-purpose plastic waste into low-cost building materials that communities used to build or expand schools.[8] This story is a great anchor point for looking at the life cycle of a plastic bottle or systems of waste management. While it's possible to study things like waste reduction and recycling in ways that do not directly connect to climate science, over-production of plastics and ineffective waste disposal both have significant climate impacts. Using compelling stories of communities finding better ways to deal with waste can serve as an entryway into understanding the complex process of producing and discarding goods that puts pressure on climate and other ecological systems.

Systems of Just Change

From day to day, we experience a tiny piece of the global system, the part that defines our local environment. And while our local place is connected to every other place on the planet, even those of us who have become attuned to this connection still filter knowledge of the world through our own, local experience. The words we use to describe the disruptions to global systems can make us feel removed from or connected to the causes, impacts, and responses to these changes.

As I discussed briefly in the first chapter, there is not a perfect term to describe the ways in which human activities have changed, and are continuing to change, global systems. You may have learned the term "global warming" when you were in school, which refers to how excess greenhouse gases in the atmosphere are resulting in an overall rise in global temperatures. However, this term focuses on only one measure of change to our global system, and since it refers to *average* global temperature over time, rather than what a particular place experiences from day to day, it can lead to confusion.

"Climate change" is a broader term that draws our attention to global climate as a system, rather than to a single measure such as temperature, and "anthropogenic climate change" specifies the changes caused by human activity. This allows us to consider how the concentration of greenhouse gases in the atmosphere is impacting not just the temperature, but as a result weather patterns across the globe. The term "climate instability" foregrounds the changes that lead to some of the most serious impacts on global systems, including humans. But these terms still aren't comprehensive. For instance, sea level rise and ocean acidification have the same primary cause but don't fit neatly into "climate."

Issues of justice are also not centered in this terminology. As we will discuss throughout this book, the impacts of climate instability are and will be greatest on communities that are already the most economically vulnerable and contribute the least to greenhouse gas emissions. For instance, people on the entire continent of Africa only contribute 2–3% of global greenhouse gas emissions, and over 500,000 people are not connected to the electrical grid at all, and yet, Africa is the continent most vulnerable to the impacts of climate change.[9] In the United States, low-income communities and communities comprised primarily of people of color are more likely to be directly exposed to the impacts of oil extraction and land degradation, and they are less likely to have access to fresh, local, and affordable food. Solutions that focus on individual action, such as homeowners paying to install rooftop solar panels or choosing to pay more for sustainably produced food, result in further protecting affluent communities from some of the impacts of climate change while increasing the impact on communities with less power in our social systems.

Understanding systems is a key to understanding the *problems* of climate change, and a systems perspective also allows us to develop just and hope-filled responses for a more livable future. When a community invests in planting trees in the parts of a city that currently have no green space, that neighborhood is better protected from the heat island effect, and the trees also help to pull greenhouse gases out of the atmosphere. When people from across a region work to set up farmers markets in places accessible to all and allow food to be purchased with Supplemental Nutrition Assistance Program (SNAP) benefits, they support local farmers transitioning to more climate-sustaining practices and promote health and well-being for their whole community. When children in a city in California learn about local efforts to create living shorelines and also learn about kelp farming efforts in small fishing towns in Maine, they begin to see how their actions are part of a global movement to re-create communities in which humans and other organisms can thrive. So while our terminology remains imperfect and incomplete, efforts to understand the *systems* impacted by anthropogenic climate change and to develop solutions that center justice in community systems moves us toward a more resilient future.

Toward Justice-Based, Hope-Filled Action

Systemic change rooted in community rather than individual action is needed to address the causes and impacts of climate instability, and community-based action allows children to see how their efforts connect to larger systems. Planting a tree in the school yard is a rewarding activity even when it is done by an individual class. But learning about and connecting with efforts to add hundreds or even thousands of trees to a region helps children see that their efforts are part of a bigger picture. Imagining how the land on which their school sits may have looked 200 or more years ago, and learning about the people who lived there and how they cared for the land allows them to connect stories and wisdom from elders to a vision for present and future communities to thrive as part of natural systems.

Considering issues of justice goes hand in hand with understanding interrelated systems. Xiye Bastida, a young climate activist who is a member of the Otomi-Toltec Indigenous Community and is now based in New York City, explains:

> Climate change doesn't discriminate, but the ability to respond to climate disasters does. Wildfires in California, for instance, are not avoiding fancy homes. Snowstorms, hailstorms, and tornadoes are not discriminating either. But fully recovering from these increasingly unnatural disasters is very difficult, almost impossible, for communities of color, Indigenous reservations, small rural towns, and traditional fishing communities.

Bastida has directly experienced the inequitable impacts of climate instability on communities based on access to wealth and power. But she also sees hope as youth rise up and demand that we consider the impacts and responses to climate change from a perspective of local and global justice. She writes: "A vibrant, fair, and regenerative future is possible—not when thousands of people do climate activism perfectly but when millions of people do the best they can."[10] As educators, we play a critical role in growing the communities of millions who are committed to "doing the best they can" rather than turning away from the challenge due to fear and feelings of powerlessness.

Children are often deeply attuned to issues of justice and fairness, and we as educators can help them develop this drive toward justice through hope-filled actions. Consider centering the stories of indigenous communities and communities of color in your choice of read-alouds, so that children see themselves in stories and also consider alternate ways of experiencing and conceptualizing the world. Invite leaders of climate justice efforts in your area to speak with your students, as they are often eager to help our youngest community members understand how they can be part of systems of repair and renewal. As we help children understand their roles as part of local, regional, and global systems, we can help them develop their identities as people who have great power as part of a community, an important antidote to the feelings of powerlessness that can overtake us when we consider ourselves as solo actors in the face of climate change.

When we consider everything we do and everything and everyone we come in contact with as connected to global systems, our perspective expands to include our community and others that we may never see in person but which are connected to us by water, air, land, and the living things that make earth such an amazing place. Supporting children to maintain their sense of wonder while learning about earth's systems allows them to stay grounded in hope and empowered action.

Justice-Based Climate Science Unit Example: Designing a Pollinator-Friendly Habitat Aligned with NGSS for Kindergarten

Guiding Question: How can we make our community a good place for pollinators? (PE K-ESS3-1)		
Focal Disciplinary Core Idea* Living things need water, air, and resources from the land, and they live in places that have the things they need. (ESS3.A)	**Focal Science and Engineering Practice** Constructing explanations and designing solutions	**Focal Cross-Cutting Concept** Systems and system models
Engage with concept and community	Observe some pollinators! This might be a long-term investigation such as caring for caterpillars as they transform into butterflies. It could also be an outdoor insect hunt where students look for an insect, point it out to a friend, draw it, and then share with the class. Short video footage (perhaps with no narration) also allows children to observe insect pollinators if they are not readily fond in your schoolyard.	

(*Continued*)

(Continued)

Explore ideas grounded in place	Learn about one or more insect pollinators that is important in your community. Use books, videos, local experts (ex/ gardeners and farmers) to understand how this insect is an important part of your community system. An ongoing classroom mural project (on a large piece of butcher paper) can help children bring their growing systems knowledge to life! They can start by drawing just the insect and gradually add the environment that it interacts with as you learn more about garden or farm systems.
Define problem in need of action	Explain to children that sometimes pollinators have a hard time getting everything they need to live. You might take a walk to a concrete-covered part of the play yard or show a photo of an empty lot and ask: what would we need to do to make this a place where pollinators can live and help plants grow? Invite children to share ideas in conversation and drawing.
Design hope-filled actions	In teams or as a whole class, design a hope-filled action. Sometimes, this needs to be fairly teacher led given implementation constraints (for instance, you may have approval to plant native plants in one particular place, or you may have supplies to make native bee house). Within those constraints, plan ways to prioritize children's ideas, for instance by having them design placement of plants or shape and exact structures of bee houses.
Share and learn from community	If children are working in partnership with a local community garden or farm, sharing might look like delivering their products (such as bee houses) and spending time learning more about this place and people. If they have created something at or near the school, such as planting native plants, they might construct informational signs and write notes to other classes about how to keep the plants and pollinators healthy. You may also want to invite families to celebrate with students and share their own experiences with pollinators and growing food or other types of plants.
Reflect and synthesize systems	The end of a unit is not the end of learning, but it is a time to reflect and synthesize. Consider asking each child to draw what they think the garden or place they've contributed to will look like when they are in the next grade, and then share with you how each part they've drawn gets what it needs to survive. This allows you to assess learning and plan next steps, gives children an authentic way to demonstrate complex understanding, and creates a product they can use to remember the project.

*Disciplinary Core Ideas (DCIs) are grade level specific, so I've written each model unit with a grade level in mind. However, the unit is flexible enough that you can likely change the content focus based on the DCIs for your grade level.

Recommended Children's Books

Archer, M. (2021). *Wonder walkers*. Nancy Paulsen Books.

Garcia, G. (2022). *We are all connected: Caring for each other and the earth* (N. J. Osorio, illus.). Skinned Knee Publishing.

Kutner, L. & Slade, S. (2014). *The soda bottle school: A true story of recycling, teamwork, and one crazy idea* (A. Darragh, illus.). Tibury House Publishers.

Larkin, S. (2019). *The thing about bees: A love letter*. Readers to Eaters.

Sayre, A. P. (2018). *Thank you, earth*. Greenwillow Books.

Mangal, M. (2021) *Jayden's Impossible garden* (K. Daley, illus.). Free Spirit Publishing.

Sayre, A. P. (2018). *Thank you, earth*. Greenwillow Books.

Sidman, J. (2021). *Hello earth! Poems to our planet* (M. A. Lora, illus.). Eerdmans Books for Young Readers.

Strauss, R. (2007). *One well: The story of water on earth* (R. Woods, illus.). Kids Can Press.

Notes

1 Strauss, R. (2007). *One well: The story of water on earth* (R. Woods, illus.). Kids Can Press.

2 Chiu, A. (2023, March 15). How California is using recent floods to prepare for future drought. *The Washington Post*. https://www.washingtonpost.com/climate-solutions/2023/03/15/california-groundwater-recharge-drought/

3 Woody, T. (2021, December 6). Meet the new climate refugee in town: Coyotes. *Bloomberg News*. https://www.bloomberg.com/news/features/2021-12-06/how-climate-change-forces-wildlife-into-cities

4 Smithsonian (n.d.) *Butterflies*. https://www.si.edu/spotlight/buginfo/butterfly

5 Crockett, Z. (2017, February 14). How one man repopulated a rare butterfly species in his backyard. *Vox*. https://www.vox.com/2016/7/6/12098122/california-pipevine-swallowtail-butterfly-population

6 van Kesteren, M. T. R., & Meeter, M. (2020). How to optimize knowledge construction in the brain. *npj Science of Learning, 5*(1), 5. https://doi.org/10.1038/s41539-020-0064-y

7 Willingham, D. T. (2004). Ask the cognitive scientist: The privileged status of story. *American Educator, 28*, 43–45.

8 Kutner, L. & Slade, S. (2014). *The soda bottle school: A true story of recycling, teamwork, and one crazy idea* (A. Darragh, illus.). Tibury House Publishers.

9 United Nations Environment Programme (n.d.). Responding to climate change. UN Environmental Programme Regional Initiatives: Africa. https://www.unep.org/regions/africa/regional-initiatives/responding-climate-change

10 Bastida, X. (2020). Calling in. In A. E. Johnson & K. K. Wilkinson (Eds.) *All we can save: Truth, courage, and solutions for the climate crisis* (pp. 3–7). One World.

3

The Carbon Cycle: Exploring Systems through Story

Grown Up Science	Hope-Filled Classroom Connections
• What is the carbon cycle and how does it impact our climate system? • How has human activity caused imbalance in the carbon cycle? • How can humans bring the carbon cycle back into balance?	• Act out the carbon cycle • Dream and draw de-carbonized solutions • Explore how big things are made of smaller things • Stories that focus on systems, cycles, and change • Stories of helpers in our community and around the world

It's a Friday in February, and my fifth-grade class' carbon cycle mini-unit has hit a lull. Two Fridays ago, students physically modeled how carbon travels through the atmosphere, using color-coded counting chips to represent carbon and dividing our room into earth's "spheres": the biosphere, atmosphere, lithosphere, and hydrosphere. For instance, when a student read a card about plants using carbon dioxide to make their food, students agreed we should move some of the chips representing carbon from the atmosphere to the biosphere. When the prompt

DOI: 10.4324/9781003393535-3

was using fossil fuels to power cars, they moved carbon from the lithosphere (since the fuel was previously underground) into the atmosphere (where carbon ends up when the fuel is burned).

In the second lesson, student groups created visual models of what we had acted out, showing how carbon moves through

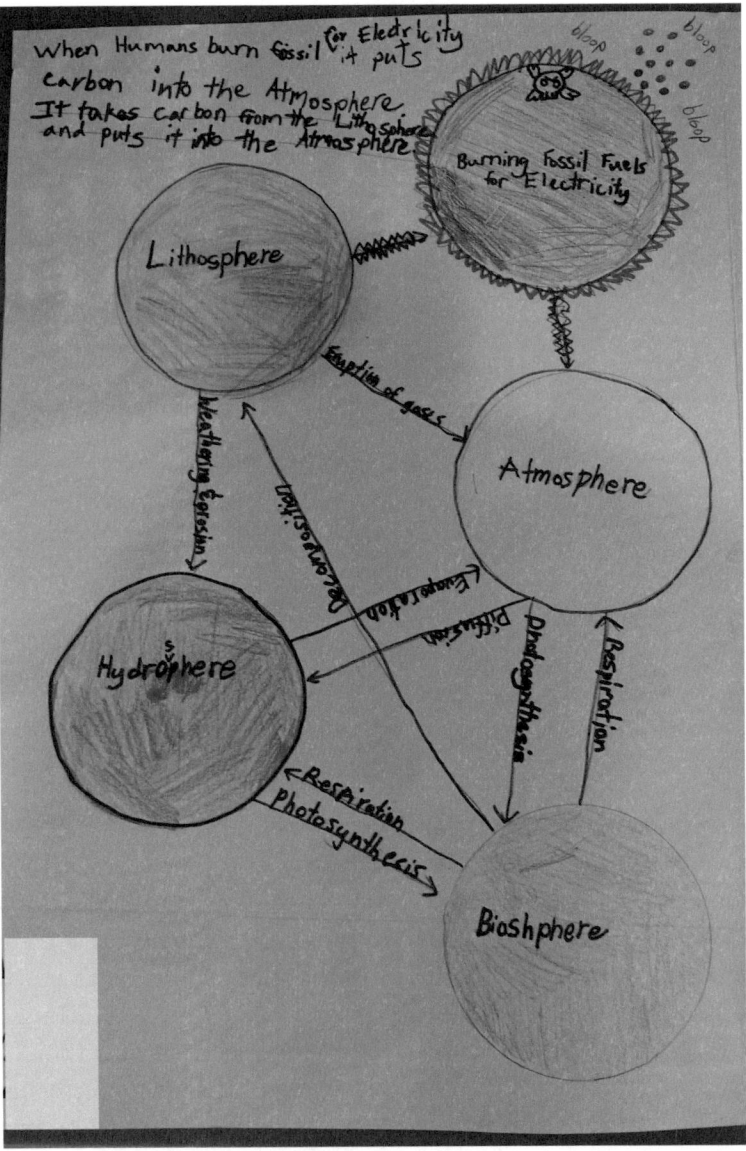

FIGURE 3.1 Students' carbon cycle diagram in progress.

earth's spheres, with each group adding a "disruptor" (such as fossil fuel burning) to show how carbon can get out of balance.

On this third Friday, we've gotten to what I hoped would be the fun part: groups imagine *positive* disruptors, actions that could help re-balance the carbon cycle by moving carbon out of the air and oceans and back into plants or the ground beneath us (including ideas to prevent carbon from being distributed to the atmosphere and ocean in the first place). Students willingly take up this task, but the ideas on every group's list looks the same: ride a bike, plant a tree, turn out the lights. All of these are, of course, great things to do! But I had hoped that groups would imagine new ideas and consider how they might be innovators. I hide my frustration, since students are, in fact, doing exactly what I asked, but I ponder how I could better support them to dream and imagine.

Then, unexpectedly, a group that had been in high conflict has a breakthrough. Tristan gets up mid-lesson and starts dancing. Luckily, I realize that it's a science-related dance before I tell him to sit back down! He is a carbon atom, or perhaps several carbon dioxide molecules, traveling from the lithosphere into a car's fuel tank and then up into the atmosphere. As he reaches the atmosphere he calls out, "oh no! I'm stuck in the atmosphere! It's getting crowded up here too." He pantomimes elbowing his carbon neighbors out of the way. As it becomes clear that the rest of the class has become an appreciative audience, his group mates join in, jumping on chairs to enter the "atmosphere" and complaining of crowds and heat. Just as I am about to step in to restore calm, Tristan turns to his partners and says, "What if it was the cars that did the photosynthesis?" Another partner says "whoah, cool!" and, perhaps aware that they have just announced their great idea to other groups, they hop down from their chairs, huddle together on the floor, and begin designing their photosynthetic car in secret.

I'm excited that this group has synthesized what we've been learning and dreamed up an innovative solution (I stifle a laugh when they note "We don't have all the details worked out quite yet") and I consider how to encourage this with all of my students. I spend the next week gathering stories of innovators who are

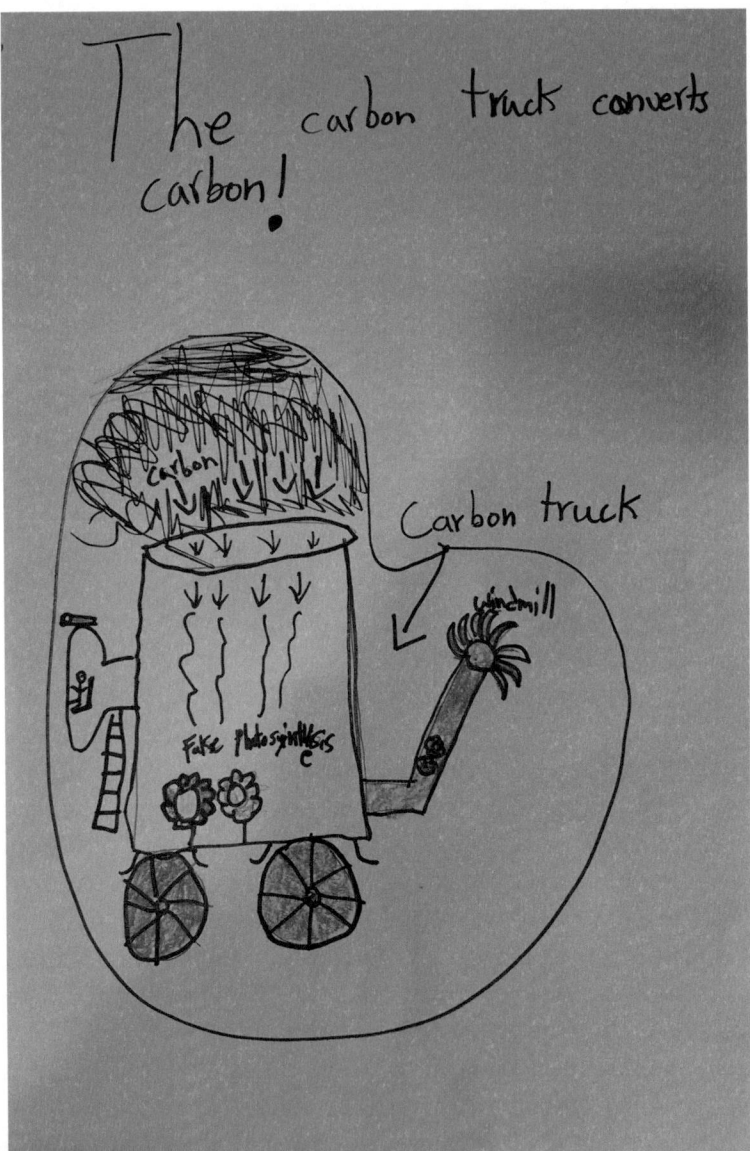

FIGURE 3.2 The photosynthetic car.

working to re-balance global carbon: kelp farmers, creators of carbon-neutral concrete, the multi-national Great Green Wall project that seeks to re-plant a huge region across much of the African continent. In our next science session, we watch short videos about these innovators, and students think about how they might be innovators too. Between the example of the photo-synthesizing car and the true stories of climate innovation, ideas begin to flow. One group proposes a greenhouse that uses plants to pull carbon out of the air and then captures the carbon they release and buries it in the ground. Several groups build on the car theme, with one drawing a car with solar panels (for day) and a windmill (for night driving) on the roof. They share their ideas in a class-wide poster session, explaining to classmates and some visiting adults how carbon has become imbalanced and why their ideas will help get things back in balance.

A Note about Teaching the Carbon Cycle in Elementary School: Systems and Stories, Not Chemical Equations

One of the main drivers of climate change is the massive, ongoing movement of carbon dioxide and other greenhouse gases into the atmosphere and ocean due to use of carbon-based energy sources. Understanding the carbon cycle and how it is being impacted by human activity is critical to addressing the climate crisis. But this is where teachers of young children sometimes get understandably worried! We may have studied the carbon cycle in high school or college, but many of us don't feel like we have a solid grasp on the chemical processes by which carbon moves from one part of the earth system to another. When we aren't solid on the science ourselves, it's difficult to see how we might make these ideas accessible to children. And the impacts of carbon imbalance are terrifying. We need ways to engage children without making them fearful.

While children are certainly capable of understanding abstract ideas when they have plenty of background knowledge, the molecular details of how matter moves through the earth's system are not a great starting point for building on children's sense of care and agency in the world around them. The NGSS do not include exploring the carbon cycle at a molecular level until high school. While the opening story shows fifth graders modeling carbon's movement through different parts of the earth's system, the goal was not understanding of molecular bonds or the details of chemical processes. Instead, it was an extension of our ongoing study of how matter and energy move through ecosystems, which is, in fact, a fifth-grade NGSS expectation. Our carbon cycle investigation came at the end of a three month study in which children had constructed miniature ecosystems, observed interactions, and developed first-hand knowledge of how energy and matter moved between parts of the system. Gradually helping children understand the world in terms of *interconnected systems* serves as a strong building block for later understanding details of the carbon cycle.

In the elementary years, students are coming to understand many systems: written language, mathematical thinking, and the social systems that govern classroom and community life. Likewise, their study of science can allow them to explore systems of how the natural world works and how humans impact and are part of these systems. Our brain is wired to make connections, and the stronger the connections between ideas and experiences, the more memorable and retrievable information is.

Throughout this chapter, the instructional techniques build on learning through story and narrative. As we discussed in Chapter 2, humans are drawn to story, and it is how cultures have built and shared knowledge throughout our history. Through hearing, engaging with, and creating story, children are able to access and connect with ideas that may seem "too hard" when presented as a series of facts. Stories are a particularly powerful way to build understanding of systems, and we'll look at several examples for children of different ages and experience levels.

Finally, the placement of this chapter—about a system that elementary children may not directly explore—near the beginning of this book has more to do with adult learning than with recommended curricular sequence. As educators seeking to engage children in hope-driven learning about our changing world, we need a solid understanding of the underlying drivers of climate change. So in this chapter, we'll use the carbon cycle as our anchoring story in order to solidify our adult knowledge of this essential system, but the suggested learning activities will focus more broadly on coming to understand systems through observation, modeling, and story.

What and Where Is Carbon?

Carbon is a chemical element that is the basis of life on earth. Living things, both plants and animals, are made of carbon-based molecules. Living things take in carbon (in the form of carbon dioxide, or CO_2) to manufacture food/ energy, and they also release carbon (dioxide) through respiration. There is a lot of carbon in the earths' system! Most of it is stored in the rocks and sediments that form the earth's crust, and a large amount of carbon is also stored deep in the ocean.

Soil, vegetation, glaciers, and the earth's atmosphere are also critically important **carbon stocks**, or places where carbon is stored. Although they hold much less of the earth's carbon than rocks do, these parts of the earth's system are becoming out of balance in terms of carbon cycling largely because human activity over the past 200 years has directly and indirectly changed the amount of carbon being stored in each of these "carbon holders."

Why does it matter how much carbon is in different parts of the earth system? When carbon bonds are broken, they release energy, which is why carbon is so very important to life. When we eat a piece of fruit, the digestive enzymes in our stomach break down the carbon-based molecules in the fruit and turn them into usable energy for our bodies (we express this energy in terms of calories). Likewise, when a pile of wood is burned,

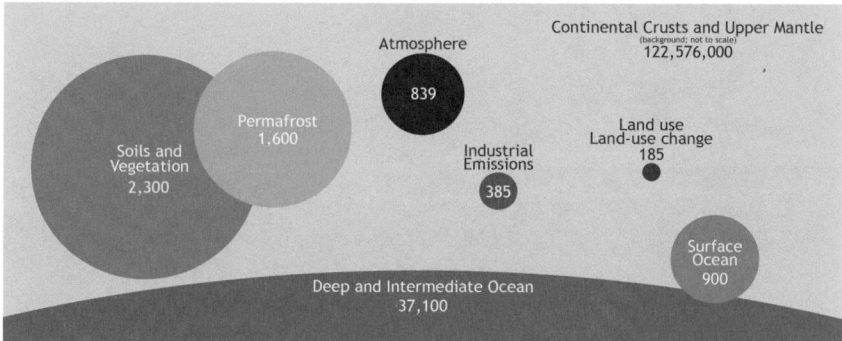

FIGURE 3.3 Global carbon stocks, shown in gigatons. Kayler, Z., Janowiak, M. & Swanston, C. Global carbon. Climate Change Resource Center. https://www.fs.usda.gov/ccrc/topics/global-carbon.

the carbon contained in what was once a tree is released in the form of heat into the atmosphere. This storing and releasing of energy powers every living thing on the planet as well as most of our human-created energy systems.

You likely know that carbon dioxide is referred to as a **greenhouse gas**. This term refers to gases in the atmosphere that absorb and emit heat. Water vapor, carbon dioxide, and methane are the most abundant greenhouse gases. Naturally occurring greenhouse gases are critical to life on earth. Without them, earth's average temperature would be below freezing! However, human activity in the past 200 years has caused a dramatic *increase* in greenhouse gases in the earth's atmosphere. The more greenhouse gases are present in the atmosphere, the more heat they trap, thus resulting in an overall rise in global temperatures.[1] We'll explore this in more detail below.

Classroom Connections: Tinier and Tinier

Young children seem to almost universally love things that are very, very tiny. On a nature walk, the adults may be admiring the sunshine, trees, and landscape, when they realize that the children have all become captivated by a trail of ants walking across the path. Building on this attention to what is often too small to be noticed by adults is a great way to set the stage for later learning about molecules.

One way to do this is to explore natural areas and substances that are made of smaller parts. For instance, a bin of soil (from a fertile garden or field, not potting soil from the store), may at first look like a pile of brown pieces. But time and a few minutes of close investigation reveals that soil is made of many different things. There are small pieces of leaves and sticks in the process of decomposing. There are tiny pebbles. Depending on where the soil is from, there may be sand or clay or even bits of shells. Putting a small scoop óf soil into water allows some of the particles to separate from each other, since some float, some sink, and others may hover in the middle of the water. Time spent exploring soil helps children become comfortable with the idea that many substances are made of tiny parts that we may not notice at first. And of course, soil is itself a system and is part of a larger system that supports life, so investigating what soil is made of can be a first step in understanding how soil is connected to plants and animals, including humans.

Looking at everyday objects under magnification is another way to build understanding that the world is made of tiny parts that don't always look the way we expect. Children love magnifiers and microscopes, although using traditional microscopes is often tricky in terms of eye hand coordination. Recent technology has made microscopy much more accessible for younger children, since projecting microscopes, including those that attach to mobile phones, can show magnified images for a whole class to explore. There are increasingly inexpensive options with clear image quality (I won't name specific ones here since the technology changes quickly). However, even without a projecting scope, children can explore the world of the microscopic. I've seen several teachers post or project a "what in the world" image of the day, an activity originated by National Geographic's children's magazines. Children look at a highly magnified image of an object that they are familiar with, but which looks very different when magnified. Human hair and skin, leaves, salt, and sugar are all fascinating to see up close! Again, this simple activity to inspire noticing and wondering helps children build their understanding of the microscopic world, so that the idea of matter made of molecules won't seem so abstract and confusing in later years.

What Is the Carbon Cycle?

This movement of carbon from one carbon stock or earth "sphere" (biosphere, lithosphere, atmosphere, hydrosphere) to another is what we call **the carbon cycle**. You may remember learning in high school chemistry that matter cannot be created or destroyed. The overall amount of carbon in the earth's system remains that same regardless of human activity. What changes is where the carbon is, and that has a huge impact on the role carbon plays in global temperature as well as ocean health.

Carbon cycles from one of earth's spheres to another in many ways. For instance, plants pull carbon dioxide from the air as one of the ingredients they use to photosynthesize (make food). This moves carbon from the atmosphere into the biosphere. When a living thing dies and decomposes, its carbon becomes part of the soil, and eventually, over millions of years, part of the rocks in the earth's crust. That means the carbon from

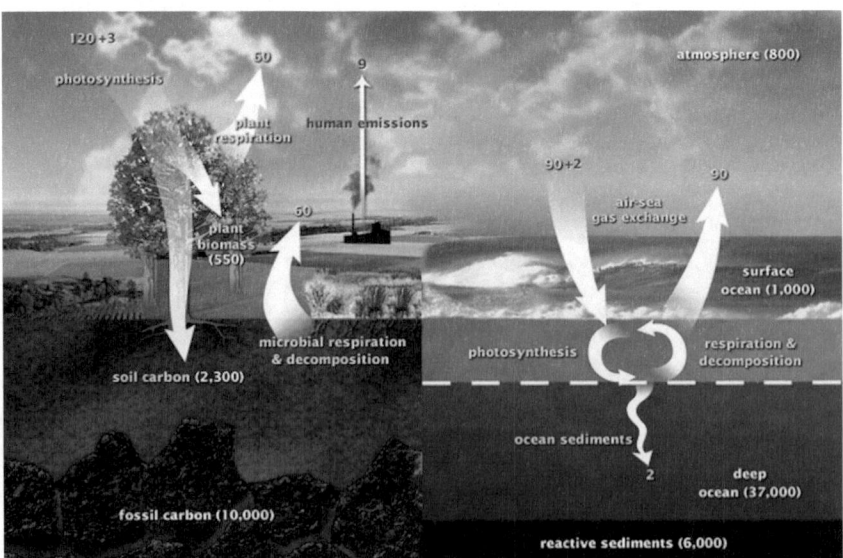

FIGURE 3.4 Movement of carbon between land, atmosphere, and oceans. NASA Earth Observatory. *The carbon cycle*. https://earthobservatory.nasa.gov/features/CarbonCycle.

the once living organism moves from the biosphere to the geosphere. This is how fossil fuels such as oil and gas are formed, through the buildup and concentration of carbon in the earth's crust over millions of years.

Carbon cycling is a natural process that occurs regardless of human activity. However, human activities have significantly changed the rate of carbon cycling and the distribution of global carbon. The biggest change comes from our extraction and use of fossil fuels from the earth's crust. Fossil fuels are valuable because they contain the stored energy from formerly living things (in the form of carbon), and this energy can be used to power human-made machines. When humans burn fossil fuels to power cars, factories, and homes, the carbon that was once deep in the earth's surface is released into the atmosphere.

While fossil fuels are by far the largest emitter of carbon-based greenhouse gases, wood-based fires also contribute to the overabundance of atmospheric carbon. Use of wood-burning methods of heating and cooking, as well as forest fires, results in the carbon that was stored in trees being released into the atmosphere. While wildfires occur as a natural part of ecosystems, direct human activity has increased the frequency and severity of these fires. As the warming atmosphere causes more unstable weather patterns, including longer and more severe droughts, this also leads to increased forest fires that both release carbon into the air and decrease the availability of trees that take carbon out of the air.

In addition to increased atmospheric CO_2, the ocean also absorbs much of the excess carbon dioxide that is released from burning of fossil fuels. This changes the pH level of the ocean water, making it more acidic. Many of the living things in the ocean are not able to thrive in more acidic water, particularly organisms that have shells or hard outer layers, including snails, crabs, and corals. The more acidic water interferes with the organism's ability to build a strong outer layer. We'll discuss the impact of climate change on oceans in more detail in a later chapter.[2]

Classroom Connections: Exploring Systems through Story

As we discussed at the beginning of this chapter, the details of the carbon cycle are important for adults to understand as they design learning experiences for children, but in the early grades, helping children understand the ideas of cycles and systems through more tangible experiences will help set them up for later understanding of this molecular cycle.

Stories of Interdependence. *The Great Kapok Tree* is a classic example of a story that helps children understand how living and non-living things are connected in systems of interdependence.[3] In the story, a man heads into a tropical forest with the intention of cutting down a very old tree. He gets tired and falls asleep, and when he does, he dreams about all of the animals who depend on the tree coming to speak to him. A bee explains that it needs the nectar from the tree's flowers for food, and in turn it pollinates the tree, allowing new ones to grow. Other animals explain how the tree's huge roots hold the soil in place, and how it provides food and shelter for many creatures. When the man wakes up, he of course decides not to chop down the tree. This story is set in a tropical rainforest, a habitat that is likely quite different from where you teach. However, the story is a great starting point for considering how plants and animals in children's local environment depend on one another. Children can use the beautiful illustrations in the book as inspiration to make art that shows all of the things that depend upon a schoolyard tree or garden or a local stream or waterway. Stories like this one that emphasize how living and non-living things are connected help build the understanding that will later allow them to understand how carbon's movement impacts so many of earth's systems.

Stories of Cycles and Change. Stories of connection help children think about systems, and so to do stories of processes. Learning the life cycle of a plant is a classic early childhood unit. A small change in focus can make this even more effective for thinking about cycles of matter in systems. Lima bean seeds are a wonderful starting point because they are easily taken apart, and the baby plant is visible inside. If children do this, and then they plant seeds, they can consider throughout the life cycle the question of "where did the seed go?" The material in the seed that surrounds the baby plant provides food before the plant is able to make its own. The seed has become part of the structure of the growing plant! As children observe the plants growing and changing, they can continue to imagine where the material that was once the seed might be now, until, eventually, flowers and then new seeds form. This repeated wondering helps build understanding that matter changes but never really goes away.

The Carbon Cycle Story. For upper elementary students, particularly fifth and sixth graders, the science curriculum often focuses on changes to matter and transfer of energy in ecosystems, and so these students may be ready to directly consider the carbon cycle. Here, too, story and narrative are powerful in helping children both understand and remember. In the opening vignette, I describe an activity in which children used small counter chips to represent carbon, and as we told short stories of natural and human-caused activities (plants using CO_2 to make food, people extracting and using fossil fuels, etc.), they moved chips to show how these activities move carbon from one part of the earth's system to another. Later, this class re-enacted this more physically, inspired by their classmate's dance representing CO_2 moving into the atmosphere. When they were working on their "hopeful action" posters, they asked if they could do a play about this for younger children in the school, and so we worked together to develop a reader's theater piece, which they acted out dramatically during a school assembly. Their finished script appears below. As with the example of *The Great Kapok Tree*, the emphasis is not on molecular details of a process but rather on a story of how living and non-living things depend on one another.

Re-Balancing the Carbon Cycle
A Readers' Theater Production
Written by Stephanie's 5th-Grade Scientists
(each line is acted out by "dancer" students in front of readers)

Reader 1: Carbon is an atom
Reader 2: A teeny, tiny piece of matter
Reader 3: A teeny tiny piece of matter that really matters.
Reader 4: All living things on Earth are made with carbon.
Readers 1, 2, 3, 4: Without carbon, there would be no life on Earth.
Reader 5: Carbon cycles through living and non-living things on Earth in a process called …
Reader 6: Wait for it …
ALL READERS: The Carbon Cycle!
Reader 6: Land plants use carbon dioxide from the AIR to make food.
Reader 7: Water plants use carbon dioxide dissolved in WATER to do the same thing.
Reader 8: Plants AND animals send carbon dioxide back OUT into the air or water.
Reader 9: Scientists call this respiration.
Reader 10: You might call it breathing.
Reader 11: When living things die, their carbon is absorbed into dirt and rocks, in the part of the earth called the lithosphere.
Reader 12: Over many years,
Reader 13: (interrupt) millions and millions and millions of years
Reader 12: that carbon can turn into oil, coal, and gas
Reader 13: We call those FOSSIL FUELS because when we use them,
Reader 14: we're using the energy from fossils.
Readers 5, 6: from the carbon of fossils.
Readers 7, 8, 9: The carbon cycle is part of how the Earth stays balanced.
Readers 10, 11, 12: Carbon cycles from living to non-living parts of the Earth
Readers 13 and 14: Over and over and over again.
ALL Readers: BUT … [dramatic pause] right now the carbon cycle is out of balance.
Reader 1: When humans burn fossil fuels for electricity, it puts carbon into the air.
Reader 2: It takes carbon from the lithosphere
Reader 3: That's the ground
Reader 2: And puts it into the atmosphere
Reader 3: That's the air.
Reader 4: When farm land gets overgrazed by cows, the soil is exposed to air.
Reader 5: Soil has carbon in it.
Reader 6: That carbon ends up washing into the water and releasing into the air.
Reader 7: To make cement for sidewalks and dams and buildings, people have to heat up rocks.
Reader 8: Heating rocks that have carbon in them releases the carbon dioxide into the air
Reader 9: When trees are cut down, they STOP taking carbon dioxide OUT of the atmosphere

Reader 10: And if they are burned, it releases the carbon back INTO the atmosphere.

Reader 12: When cars burn fossil fuels, carbon that belongs in the lithosphere …

Reader 13: … ends up in the atmosphere instead.

Readers 1, 2, 3, 4, 5, 6, 7: Right now, there is too much carbon in the atmosphere.

Readers 8, 9, 10, 11, 12, 13, 14: We need a plan to put it back where it belongs.

Reader 1: Did you ever think about growing a seaweed farm?

Readers 2: Why would I want to do that? WHERE would I even do that?

Reader 3: Seaweed can suck carbon dioxide out of the air and water and use it to make food for the plants.

Reader 1: It's called photosynthesis

Reader 2: I can't wait to be an ocean farmer!

Reader 4: If farmers plant cover crops,

Reader 5: When they aren't growing other things

Reader 6: the soil isn't exposed,

Reader 4: and the plants take carbon dioxide from the air.

Reader 7: Some scientists have developed cement that ABSORBS carbon dioxide.

Reader 8: We can use this special cement to make dams and even sidewalks.

Reader 11: Planting rooftop gardens helps take carbon out of the air.

Reader 9: The plants use carbon dioxide

Reader 10: AND they cool buildings

Reader 11: and even whole cities

Reader 10: with less electricity.

Reader 14: We designed a carbon truck that makes fake photosynthesis and uses it for energy.

Readers 1, 2, 3, 4, 5: These are just a few ideas we had for getting the carbon cycle back in balance.

Readers 6, 7, 8, 9, 10: We are the scientists and engineers of the future.

Readers 11, 12, 13, 14: But we can start working on our ideas now.

ALL Readers: What about you? What great ideas do YOU have???

What Can Be Done to Re-Balance the Carbon Cycle?

The dramatic, ongoing increase in atmospheric carbon due to human's increasing use of carbon-based energy sources is causing many changes to the earth's system, and you have likely experienced some of them directly. As described earlier, carbon in the atmosphere helps control the earth's temperature by absorbing energy emitted from the earth's surface and then reflecting it back. While this feature is critical to life on earth, the rapidly increased amount of CO_2 in the atmosphere is causing global temperatures to rise more quickly than earth systems that support life can readily adapt.

The rapid (in geological history terms) rise in global temperature is causing direct impacts in the form of increased heat waves, melting ice sheets, and sea level to rise as well as indirect impacts in the form of climate instability. As mentioned earlier, the increased temperature holds more water vapor in the atmosphere, and this in turn leads to more severe and unpredictable weather events. As the atmosphere destabilizes, there is an increase in not only in heat waves, but also in hurricanes, droughts, and other extreme weather events.

The impacts of an imbalanced carbon cycle are enormous, and thinking about them can cause us to become hopeless. Without hope, people have difficulty engaging in problem-solving or staying engaged with a problem, even when they know it is critically important. Recent research has shown that young people's mental health is being negatively impacted by worry over climate change and society's collective failure to act.[4] So when we work on ideas related to the carbon cycle with children, I recommend we NOT dwell on the huge magnitude of the problem, potentially causing more distress. Instead, once children understand the basic narrative of the carbon cycle and how it is impacted by humans, we can turn our attention to stories of hope-filled action in our own communities and in collective actions around the world.

At a national and global level, de-carbonization of our energy systems is one of the most critical parts of reducing climate instability. This means replacing carbon-based forms of energy production—coal, oil, and natural gas—with systems that do not result in carbon being released into the atmosphere, for example, solar, wind, and nuclear (although the latter, while carbon-neutral, is controversial for other reasons). In addition to slowing or halting the addition of more and more carbon to the atmosphere, we also need systems to reduce what is already there, in order to re-stabilize global climate.

As of this writing, there has been far too little progress in these areas in terms of national policy. There are, however, many points of hope. Groups big and small are engaged in collective action both to develop new, carbon-neutral and carbon-capture technologies and to implement solutions that we already have the knowledge and tools to enact. To create an environment in which learners feel that they have agency to make a positive difference in this global challenge, exploring the stories of hope-in-action is critical.

While we want to help children explore the big picture systems that connect different parts of the earth, when it comes to stories of action, often small and local is better. Advocating for changes to how our power is generated is important, but so is planting and tending to a small garden or caring for the street trees near a school building, and the latter activities allow children to directly see the impact of their actions. Supporting children to feel the power of their small actions helps them grow into adults attuned to and ready to take on larger scale change. The final section of this chapter provides examples of climate action stories that have inspired and empowered my students.

Hope-Filled Action: Stories of Hope to Spark Imagination and Innovation

When we tell stories of people and communities who work to address climate-related issues, rather than building a sense of dread at the state of the world, we help build a sense of agency and power. Stories of changemakers encourage children to imagine and design their own solutions even if, like the photosynthetic car that my students envisioned, imagination is a little ahead of implementation. Dreaming what is possible is something children are great at, and encouraging this leads us toward a more hopeful future. Here are a few stories of hope that have resonated particularly powerfully with my students over the years. They will come up again in chapters about specific systems impacted by climate instability. I encourage you to use these as inspiration to explore your community and consider the interests of your students in finding stories of hope-filled action to share with your students.

Vertical Forests and Green Roofs. In several cities around the world, architects and engineers have worked together to design buildings that are also living ecosystems. Architect and urban planner Stefano Boeri designed the first prototype of this type of building/ forest in Milan, Italy, and the idea is being further developed around the world. Children and adults alike are excited and inspired by photos of these building/ forest combinations, and it becomes easy to imagine ways to make our buildings quite literally more green.[5] A more well-known and widely used application of a similar idea is the green roof, where traditional roofing material is replaced by plants, which provide insulation for the building as well as helping reduce the increased heat that urban areas experience, while also adding photosynthesizing (and thus carbon-capturing) plants to environments where plants are otherwise scarce.[6] You may be able to find examples of green roofs in your own community, and versions of them have been used in different areas around the world, so exploring green roofs provides an opportunity to learn stories of communities near and far.

Kelp Farming. Although I live and teach in a community adjacent to a large bay and the Pacific Ocean, I've not taught in schools where most of my students' families are directly connected to the water through their work or everyday lives. So I was surprised at how excited my students have gotten when introduced to the stories of kelp farmers, who grow seaweed as both an economic and environmental tool. Kelp is used as a food product and also has a number of agricultural and industrial uses, so there is economic incentive to grow and harvest kelp.[7] Kelp farms, just like forests on land, are powerful carbon sinks, drawing carbon out of the air due to photosynthesis. Kelp grows much faster than most land plants, so it is able to extract CO_2 out of the air at a rapid rate.[8] Kelp farms also help reduce ocean acidification, as it removes nitrogen and phosphorous from the water.[9] In addition to kelp farming, there are a number of projects investigating how to use kelp on a large scale to both capture and sequester atmospheric carbon. The story of Running Tide Technologies, which is working to engineer structures that will grow kelp and then sink into the ocean to keep the captured carbon deep undersea, is a great, accessible example of environmental engineering that may inspire children to develop more creative solutions.[10] Learning about kelp farming has led my students to dream about many other ways we might use the power of photosynthesis to help re-balance the carbon cycle.

Justice-Based Solar Transition. Rooftop solar is one of many ways to transition away from fossil fuels as the energy source for homes, businesses, and industry. However, relying on individual homeowners to install solar panels creates and grows inequity, as those who are affluent are able to transition to and benefit from clean and renewable energy, while renters and those with fewer resources are unable to make this change. In Minneapolis, the group Minneapolis Climate Action has collaborated with other community groups and businesses to develop "solar gardens," installations that serve whole neighborhoods, including renters, rather than individual homeowners. One of these is located on the large, flat roof of a local school. They use a subscription model that results in reduced electrical bills for community members.[11] Stories of collective action that benefits whole communities show what a "just transition" looks like, an idea we'll take up again in Chapter 8.

The Great Green Wall and other Reforestation Projects. It would be hard not to be awed and inspired by the communities who are working to conceptualize and build the Great Green Wall, a massive effort across the Sahel region in Africa, using indigenous knowledge and planting techniques to restore previously plant-rich land that has become desert due to drought, climate change, and unsustainable farming practices. While the Great Green Wall was originally envisioned as a massive tree-planting effort across the continent, it has transformed into an effort that is better informed by the farmers who have lived on and cared for this land for generations. The ongoing story of the Great Green Wall is a wonderful story of someone having a good idea that turned out not to be quite right (planting a massive forest across an arid region), and then listening to and working with communities to transform the idea into something more effective.[12]

Up Next: Place-Based Learning for Hopeful Action

As I explained at the start of the chapter, understanding the carbon cycle is essential for educators who are integrating climate education into their instruction, but it is not the best concrete starting point for climate-based explorations with young children. The chapters that follow explore specific natural and human-constructed environments that allow children to build understanding of interconnected systems and take hopeful action in the world around them.

**Justice-Based Climate Science Unit Example:
Re-Balancing the Carbon Cycle
Aligned with NGSS for Grade 5**

Guiding Question: How can humans work together to bring the carbon cycle back into balance? (5-ESS2-1, 5-ESS3-1)		
Focal Disciplinary Core Idea Earth's major systems are the geosphere, the hydrosphere, the atmosphere, and the biosphere. These systems interact in multiple ways to affect earth's surface materials and processes. (ESS2.A)	**Focal Science and Engineering Practices** Developing and using models Constructing explanations and designing solutions	**Focal Cross-Cutting Concepts** Systems and system models Scale, proportion, and quantity

(Continued)

Note 1: This unit works best after students have explored energy transfer in ecosystems (5-PS3-3, 5-LS1-1, 5-LS2-1) and are familiar with the exchange of matter and energy between the atmosphere, plants, and animals.

Note 2: These lessons are adapted from the California Academy of Science's Instructional Resources on the carbon cycle. I highly recommend referring to them directly if you plan to teach this unit.[13]

Engage with concept and community	Carbon Cycle Role-Play Part 1 (living systems): Because understanding the carbon cycle involves matter that we cannot see (carbon dioxide and other gases), physical modeling is a helpful support. 1. Put students into seven groups. It's helpful for them to wear colorful paper headbands with the name of their group written on them so they are easy to track as they move around. Groups represent: atmosphere; ocean, lakes, and rivers (hydrosphere); rocks and soil (geosphere); land plants (biosphere); water plants (biosphere); land plants (biosphere); land animals (biosphere). Provide each group with an information sheet about their sphere and ask them to brainstorm and post a list of examples of things in their sphere that as it looks in their local community. 2. Have each group introduce their sphere (or part of a sphere to the whole class). 3. Review (or introduce) what carbon is and have students share what they know. It is helpful to provide visuals so that they can consider carbon in many different forms, including carbon dioxide. Then give each group a clear cup of eight linker cubes representing carbon. Review that these cubes are not actually carbon! They will help us model and visualize a process that is hard to see and takes place over long time periods. 4. Using examples from the Cal Academy of Sciences lesson plan as well as examples students develop, have students act out the carbon cycle by physically giving carbon from one sphere to another as you call out an example. For example, if you read "trees use carbon dioxide from the for photosynthesis, so they can convert sunlight into food energy," the atmosphere group would bring some of its carbon to the land plants. Ask them to verbalize their action, such as "we're giving carbon dioxide gas to the plants. They'll use it to make their food." 5. I have found that it is helpful to spread out this role-play over at least two different lessons to give students time to reflect and synthesize their ideas. Repeating some of the same prompts as well as adding new ones helps students build confidence and understanding.

(Continued)

(Continued)

Explore ideas grounded in place	1. Have small groups create visual representations of how carbon moves through the four global spheres. A sample guideline sheet is included at the end of this outline. Once groups have created their representations, they can share with each other and discuss how they have chosen to represent the different parts of the cycle. 2. In a separate lesson, post a set of "disruptor" scenarios (choose from among those provided in the Cal Academy lesson and/ or develop examples in collaboration with students). Discuss one disruptor in terms of which part of the carbon cycle it disrupts and how. Then let groups choose a few to add to their posters. It works well to have students draw these on separate, smaller paper so they can be attached and removed from their carbon cycle posters
Define problem in need of action	Spend time learning about "carbon heroes" in your community and beyond. You might do this by watching short videos and news segments on innovations such as: algae farming, reforestation projects, solar energy projects, vertical forests, urban bikeways, carbon-negative concrete, and other efforts specific to your community. Students can add to a growing table to record who the carbon heroes are, how their work helps re-balance the carbon cycle, what part of the cycle it impacts, and ideas they have after learning about the example.
Design hope-filled actions	Small groups can now work to design their own carbon re-balancing innovation. Encourage groups to brainstorm a list of ideas before they decide on one that they will illustrate and share. Remind them that they can dream up innovations that don't already exist and that dreaming of what could be is an important part of making the world a better place! Once each group has decided on an innovation, ask them to "dream and draw" to create a second poster that shows what their idea is and how it will help their community to re-balance the carbon cycle.
Share and learn from community	Students can hold a poster conference in which they take turns being listeners and presenters. This helps develop oral communication skills and treats their ideas as serious and worthy of sharing. Consider inviting parents, community members, or students from other classrooms to take part in the sharing.
Reflect and synthesize systems	Ask students to reflect on what they have learned from each other and from the examples of "carbon heroes" that they have learned about during the unit. You may want them to consider the question of "what next?" What steps would they like to take as a class to advocate for change in their community, to help move toward a future in which the world's carbon cycle is back in balance?

Carbon Cycle Poster Guideline

- [ ] First, you need to represent each of earth's "spheres." You can use colors and/ or shapes to represent these. Remember you will be connecting them with arrows, so your design needs to allow for that.
- [ ] Be sure to label each sphere with its name AND have visuals to help us remember that the sphere is. Use visuals that will help your classmates remember what this sphere looks like where we live.
- [ ] Now draw and label arrows that show the flow of carbon TO and FROM each sphere. See the "ways carbon moves" list below for words you may use to label your arrows.
- [ ] When you are done, have another group check your work for completeness and readability.
- [ ] Ask your teacher to give you a disruptor card. Discuss how you think the disruptor impacts the carbon cycle. When you feel confident in your ideas, check in with your teacher.
- [ ] Design a way to represent the disruptor on your poster. You want to make it stand out from everything that's already there!

Next week you will add on a part with your ideas to disrupt the disruptor and put the carbon cycle back in balance!

Earth's "Spheres"	Ways carbon moves
Lithosphere Atmosphere Hydrosphere Biosphere	Photosynthesis Respiration Decomposition Eruption of gases Weathering and erosion Evaporation Diffusion Dissolving

Recommended Children's Books

Bang, M. (2009). *Living sunlight: How plants bring the earth to life*. The Blue Sky Press.

Cole, J. (2010). *The magic school and the climate challenge* (B. Degen, illus.). Scholastic.

Kirby, L. (2021). *Old enough to save the planet* (A. Lirius, illus.). Harry N. Abrams.

Lyon, G. E. (2011). *All the water in the world* (K. Tillotson, illus.). Atheneum/ Richard Jackson Books.

Notes

1 National Aeronautics and Space Administration (2023, April 5). *Quiz: Global warming*. https://climate.nasa.gov/climate_resources/16/quiz-global-warming/

2 National Oceanic and Atmospheric Administration (2019, February 1). *Carbon cycle*. https://www.noaa.gov/education/resource-collections/climate/carbon-cycle

3 Cherry, L. (1990). *The Great Kapok Tree*. San Diego: Harcourt.

4 Thompson, T. (2021). Young people's climate anxiety revealed in landmark survey. *Nature, 597*(7878), 605–605. https://www.nature.com/articles/d41586-021-02582-8

5 Xie, J. (2017, August 9). *Vertical forests may help solve climate change and housing shortages*. Curbed. https://archive.curbed.com/2017/8/9/16059384/vertical-forest-italy-climate-change

6 Environmental Protection Agency (2022, July 13). *Using green roofs to prevent heat islands*. https://www.epa.gov/heatislands/using-green-roofs-reduce-heat-islands

7 Bailey, A. (2021, December 3). *Kelp farming shows promise as new industry*. University of Alaska, Fairbanks. https://uaf.edu/news/kelp-farming-shows-promise-as-new-industry.php

8 Hurlimann, S. (2019, July 4). How kelp naturally combats global climate change. *Science in the News*. https://sitn.hms.harvard.edu/flash/2019/how-kelp-naturally-combats-global-climate-change/

9 Suffolk county pilot program touts potential benefits of harvesting kelp. (2017, March 7). *CBS New York* https://www.cbsnews.com/newyork/news/long-island-kelp-farms/

10 Bever, F. (2021, March 1). 'Run the oil industry in reverse': Fighting climate change by farming kelp. *National Public Radio*. https://www.npr.org/2021/03/01/970670565/run-the-oil-industry-in-reverse-fighting-climate-change-by-farming-kelp

11 Minneapolis Climate Action. (n.d.) *Community solar and environmental justice*. https://www.mplsclimate.org/community-solar-and-environmental-justice.html

12 Morrison, J. (2016, August 23). The "great green wall" didn't stop desertification, but it evolved into something that might. *Smithso-*

nian Magazine. https://www.smithsonianmag.com/science-nature/great-green-wall-stop-desertification-not-so-much-180960171/

13 California Academy of Sciences Curricular Resources on teaching the carbon cycle: https://www.calacademy.org/educators/lesson-plans/carbon-cycle-role-play and https://www.calacademy.org/educators/lesson-plans/carbon-cycle-poster

4

Getting to Know Trees and Forest Systems

Grown Up Science	Hope-Filled Classroom Connections
• How are forests interconnected systems? • How do trees and other plants cycle gases in and out of the atmosphere? • How are forests essential to the carbon cycle and climate system stability? • Why are healthy forest ecosystems important even if you don't live in or near a forest?	• Explore and wonder in a forest • Observe a tree in your community over time • Develop understanding of photosynthesis through story and investigation • Advocate for tree justice in your community • Explore forests as systems: looking beyond the trees • Stories of hope-filled action to protect and grow the world's forests

We are walking through the forest, or more accurately, running, skipping, and coming to sudden stops. We arrive at a small clearing, and there is a pause in the playful energy as the children look up at the surrounding trees, taking in their size, comparing it to their own, now small-seeming stature. Sirena walks over to one and wraps her arms around it, looking up as she does so. "It's like hugging a giant. But prickly" she says to no one

DOI: 10.4324/9781003393535-4

in particular. A group of her friends gather around to also hug the tree. They report on the smell, the texture, the way stands without moving despite being surrounded by first graders. One student gives the tree a name and begins talking to it as if it is a very tall grown-up standing in their path.

We are on a field trip to a regional park just a few minutes from our school, and the first graders are excited to be outside in the middle of a school day. It is December, but only mildly cold, and the predicted rain has held off. The naturalist who was supposed to meet us for our walk was not able to join us, so the classroom teacher, a classroom parent who is knowledgeable about local plants, and I are with the children, trying to respond to their noticings and provide some information without being too directive. When Sirena and others begin hugging their new tree friend, I see an opening. "Wow, that tree is big," I agree. I ask them if they can hug all the way around it, and they quickly tell me they can't, although a couple of them try to stretch their arms longer. Then Lana grabs Rosie's hand and together they are able to wrap their arms around and touch fingertips on the other side. "It's a two-person tree," Lana reports. "I wonder how they got so big," I say, but the children have now become interested in how many people it takes to circle each tree. They find one that needs three kids to stretch around, and they search for an even larger one. They also quickly realize that their arms are of varying lengths, and that Kyla and Tyler, both of whom are quite tall, can hug the circumference of some trees that take three other students. Lana, who has taken charge of this exploration, announces that a tree will "count" as a four-person tree as long as it takes four of the smallest students to wrap their arms around. The trees in this part of the forest aren't huge, and they struggle to find one that is as big as they are hoping. Some children try to shorten their arms by bending them to convince themselves that they have found a four-person tree.

Later, as the children have a snack, the parent chaperone asks them if they think all the trees they found are the same kind or different. Sam says different, since some were skinny and some were fat. Jordan looks down at the ground and says, "the leaves look all the same. I think they're the same, but maybe some are babies and some are the parents." This is, in fact, the case.

I try my question again, "I wonder how trees get big like that?" Several children share ideas, mostly making connections to how they themselves grow. One student says that the trees eat up the soil and get bigger and stronger. Another wonders if trees have mouths, and they start to giggle about trees talking to each other while they snack on soil. These moments are tricky for me as a science educator. Trees don't eat soil, and this is a common misconception. I want the children to start to understand how plants are fundamentally different than animals in terms of how they process energy. But I don't want to cut off the wondering and meaning-making that is happening.

I decide not to directly address the soil-eating idea right now. Instead, I say, "you know, trees, and all plants, have a superpower that we don't have. They can make their own food inside of them! Did anybody ever hear that before?" Several children give exaggerated "WHAT?" responses and laugh. Rosie says "My mom told me plants breathe but it's like the opposite of people." I say, "that's part of how they make their food! If we ate air, would that make us grow big?" Children laugh at this idea and agree that they would be really hungry. I say "plants don't really eat air either, but they use it to make their food." I briefly explain that the sun is full of energy, and we can see and feel the sun's energy on earth. I tell them that plants have a special chemical in them that helps them use water and air to turn the sunlight energy into food they can use to grow. Laurel says, "cause if you put a plant with no light it will die." Others nod in agreement and look up at the cloud covered sky. Sam yells upward, "hey sun, these trees are hungry!" Everyone laughs. I decide I've given enough direct instruction for the moment, and the children go back to their wide-ranging conversations.

A few days later, back in the classroom, we return to our discussion. I have posted a large paper cut out of a tree on the white board, and we're sitting in a semi-circle facing it. I ask if anyone would like to pretend to be the air. Nearly everyone waves eager hands. I assure them everyone will get a turn and ask four students to practice "floating" around our circle area to represent the air that is surrounding the tree. Then I ask four other students to pretend to be the water that is in the soil. They bend down below the tree cut out and pretend they are being sucked up by the tree's roots. The children think that all of this is hilarious, but they stay hard at work at their jobs. I then explain that some water makes it all the way up to the leaves of the tree, and that the air got into the leaves through tiny holes. The actors gather in the middle of the circle and pretend to mix together inside an imaginary leaf. I say,

> Imagine it's a sunny day. The sun's light is energy, and it's shining on the leaf. There is a special chemical in the leaves that can do something amazing. It can use the sunlight to make part of the water and part of the air combine, and then they can hold some of that energy.

I have each water actor hold the hand of an air actor, and I give them a yellow pom pom to hold in their clasped hands, representing the energy held between the molecular bonds of glucose (I don't use those words, just "energy"). I say "believe it or not, those leaves turn water and air into a kind of sugar that plants use to grow. Your friends are plant sugar now!" The group expresses a desire to eat their classmates. I respond, "sometimes the sugar does turn into things people can eat, like fruit. But the plant also uses the sugar to make all its parts, like even the tree trunk." Before this can get too out of control, I thank the actors for their help and ask them to return to the circle. For several more minutes, other students take turns pretending to be air and water. As we do this, I repeatedly mention that the leaves are turning energy from the sunlight into food that the plant can use to grow. After about 15 minutes of acting out this very simplified version of photosynthesis, I ask the students to draw and write in their science notebooks to show how the tree gets the food it needs to grow bigger and bigger.

In future discussion circles, we discuss, and again act out, the idea that the leaves have leftovers when they turn water and air into food/energy for the plant. The leftovers include oxygen, which animals need to stay alive. A group of students become obsessed with the idea that "trees poop oxygen," and while I don't repeat or reinforce this not-completely-accurate idea, I suspect that those children will never forget that oxygen is a byproduct of photosynthesis! With their main classroom teacher, the students read several books about trees. In our final science discussion before winter break, they share with me that they've learned that trees help "take the bad stuff out of the air" and turn it into healthy oxygen. Again, this understanding is a simplification, in that carbon dioxide is not "bad" and is in fact necessary for life on earth. But they are beginning to understand how trees are part of the system of gas exchange in the atmosphere, and that they play an important role in keeping atmospheric gases in balance.

In the spring, I will work with this class as they study pollinators including butterflies and bees, and they will build a pollinator garden in a small patch of soil at the edge of the school grounds (see Chapter 2 for this story). This provides an opportunity to return to the idea of how plants, animals, people, and the non-living environment are all connected and dependent on one another. For now, I am excited that this group of children is so eager to talk about trees and the fascinating chemistry of how they, and all plants, are part of the earth system.

Forests as Connected Systems

When was the last time you walked through a forest, big or small? What did it feel like to be surrounded by trees? What smells and sounds did you notice? I grew up in the Appalachian Mountains, and spending time in forests was part of my daily life that I took for granted. I've lived in urban areas my whole adult life, and on the West Coast for much of it, but sometimes I smell something in the air that transports me back to the deciduous forests of Virginia, the smell of slowly decaying leaf litter in fall or of new, green growth in spring. There has been some recent interest in the US media about the Japanese practice of "forest bathing," and when I first read about this, I understood immediately. The Appalachian forests still feel like a part of who I am, and there is something about being surrounded by huge trees that helps me understand that I am part of a system much larger than myself.

Forest systems are essential not just to the happiness of those who grew up near them, but to the health and stability of our planet.

Forests are complex ecosystems, and the trees that define this system are inextricably tied to all the living and non-living things around them. They provide food and shelter for a wide range of animals, from tiny insects to apex predators. Other plants grow in the shade of the forest floor and even on the trees themselves, as in the case of some mosses and epiphytes. The trees in a forest process huge amounts of carbon as they photosynthesize, transforming energy from sunlight into the food needed to grow and survive. And they require resources from the environment around them, most notably water, a critical part of the photosynthesis process.

Forests cover over 30% of the land on earth, in environments ranging from dense, tropical rainforests to the deciduous, temperate forests of my childhood, to the boreal forests that cover much of the land in northern latitudes.[1] Although the species of plants and animals that live in each of these environments differ greatly, as does the climate, soil, and other non-living features, all forests are complex systems of living and non-living things that play an integral role in both global climate and biodiversity.

In this chapter, we will explore the role that forest systems play in climate regulation and identify ways that scientists and climate activists are working to build, protect, and sustain forest environments as part of efforts to reduce climate instability. While the informational part of this chapter focuses primarily on the biochemistry of plants and their impact on climate, the suggested engagements with children will include exploration of forest biodiversity, as this is also essential to planet health and is often an accessible entry point for children to explore complex environments.

Classroom Connections: Get to Know a Tree

How might we introduce children to understanding forest systems? Part of this answer will depend on the environment in which you live. If your community is in or adjacent to forested areas, then direct exploration of a forest system is the most powerful approach to building both knowledge and personal connection. But even in areas that are not forested, there are trees, and these trees are part of more complex systems. So I recommend beginning with what is locally accessible and meaningful before considering the study of forests around the world. At any grade level from preschool through adulthood, **studying a single tree over the course of a year** is fascinating. Many teachers have tried this by choosing a tree that is close to or on school grounds and asking their students to get to know the tree and engage in more and more nuanced observations over time.

You might begin by just sitting around the tree quietly, noticing what it feels like to look up and see the tree canopy overhead. What do the leaves look like? What do they feel like? What about the texture of the bark on the trunk? How big is the tree? Young children can measure in terms of arms wrapped around the tree, as my first graders did in our opening story, and they might estimate the height in terms of how many first graders tall they think it is. Older children can use a tape measure to measure the circumference and note if there is any change over the course of a year. With a little help they can even estimate the height of the tree by comparing the size of its shadow to the shadow size of an object of known height, like a meter stick.[2]

Over the course of a school year, the tree will change in ways that provide openings for exploring concepts such as photosynthesis and respiration, complex processes that children can begin to understand conceptually long before they need to formally balance a chemical equation. As children have ample time to observe and get to know this local tree, they will also become more aware of the other living things that interact with and depend on the tree. They may see birds resting on branches or even building nests. Insects may crawl up the trunk or spend time at the base of the tree. They may find worms in the rich soil formed by decaying leaf litter. Keeping a journal of all they observe and using this as a basis for further exploration help build an understanding of how trees are a critical part of ecological systems.

This simple act of taking time to regularly observe and think about a local tree can form the building blocks for all three of the core components that center this book: understanding systems, coming to know and understand the place of which we're a part, and engaging in hope-filled action to preserve and increase these amazing living things that help keep earth's systems in balance.

How Trees Impact Carbon Distribution: A Brief Review of Biochemistry

We likely all remember learning, at some point in our education, that "trees [and other plants] produce oxygen." Perhaps you've also heard forests referred to as the "lungs of the planet." Both of these statements reveal important aspects of the role that forests play in the interconnected systems that allow for life on earth. Let's go a bit deeper and review some of the essential biochemistry that results in forests playing such a large role in the air we breathe and in global climate systems. Please note that the explanation I provide here is somewhat simplified. The biochemical processes in plants are fascinating, but they are also complex. This section provides a broad overview of adult-level background knowledge to inform design of learning experiences for children. If the topic is interesting to you, and if it has been a while since you took a plant biology class, I encourage you to look beyond this book for more detailed and complete explanations of how plants produce and use energy-storing molecules.

All of the energy that allows living things on earth to grow, develop, and reproduce comes from the sun. Plants are *producers*, meaning that they directly convert energy from sunlight into energy that they can use for growth and reproduction. Animals can't do that! Animals, including humans, must consume food as our source of energy, which is why animals are called *consumers*. So how do producers "produce" enough energy to create a giant tree? Through the magical-seeming twin processes of photosynthesis and respiration.

Photosynthesis happens inside the leaves of plants. You may recall that one way plant cells are different from animal cells is the presence of *chloroplasts*. Chloroplasts are a structure inside the cell that contains a chemical called chlorophyll. Chlorophyll is a pigment, which means a molecule that absorbs light at a specific wavelength. Chlorophyll absorbs light energy in a way that allows the plant to convert the energy from sunlight into stored energy in the form of glucose (a type of simple sugar).[3]

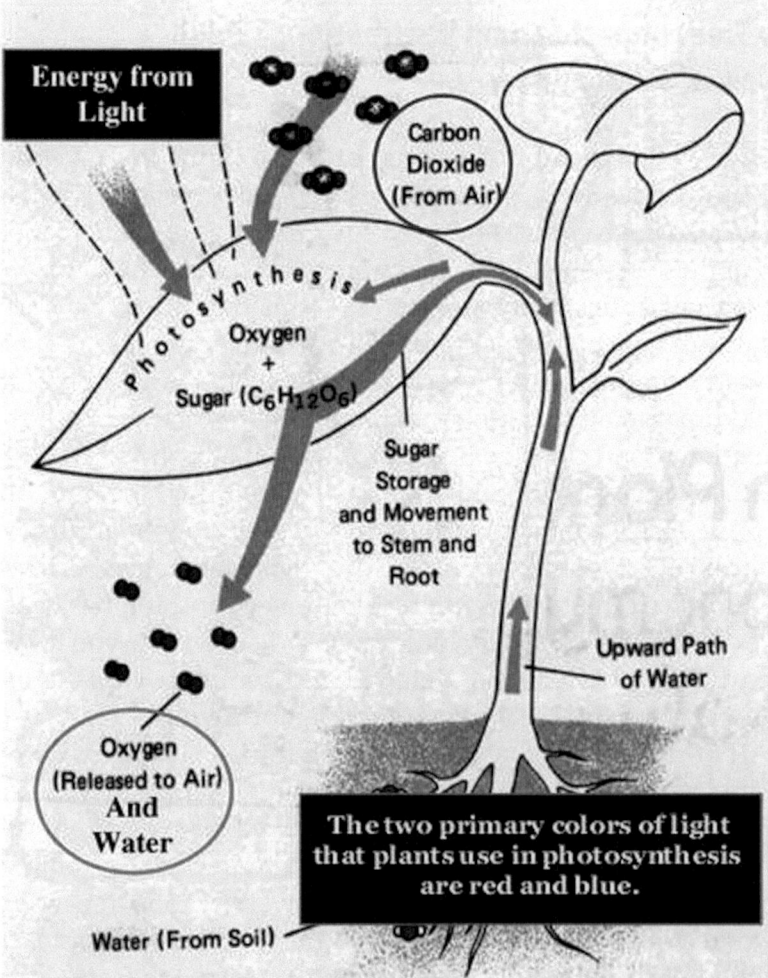

FIGURE 4.1 Photosynthesis process. NASA (2007, April 11). NASA predicts non-green plants on other planets. https://www.nasa.gov/centers/goddard/news/topstory/2007/spectrum_plants.html.

The ingredients of photosynthesis are water and carbon dioxide. Plants soak up water, usually from their roots, and transport it to the leaves. Carbon dioxide enters the leaves through tiny holes called stomata, generally found on the underside of leaves (this prevents too much evaporation of the water, which would be more rapid if stomata were on the tops of leaves and more exposed to the sun). Using the energy from sunlight, a chemical reaction happens inside the leaf that turns water and carbon dioxide into glucose, which stores energy for the plant, and oxygen, which is what is left over from the chemical reaction.

I've found that older elementary students are excited to learn part of the language of chemistry in coming to understand photosynthesis. Many of them already know that water is also referred to as "H two O" which is written as H_2O. That means that a water molecule is made up of two hydrogen atoms and one oxygen atom. When I teach the basics of photosynthesis with fifth graders, we use different colors of linker cubes to represent the hydrogen and the oxygen. Note that the linker cube model does not accurately show relative size or the nature of molecular bonds, but that's not the goal at this level.

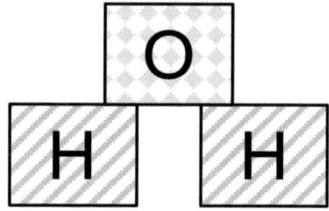

FIGURE 4.2

Carbon Dioxide is written as CO_2, meaning that it is made up of one carbon atom and two oxygen atoms.

FIGURE 4.3

Using the cube representation, here's a model of the chemical compound glucose, the sugar that is produced from photosynthesis.

H	H	H	H	H	
H	O	O	O	O	O	O
	C	C	C	C	C	C
	H	H	H	H	H	H

FIGURE 4.4

So glucose is made of six carbon atoms, 12 hydrogen atoms, and six oxygen atoms and is written as $C_6H_{12}O_6$. How is it possible for water and carbon dioxide to create this larger molecule? You can model this by figuring out how many water molecules and how many carbon dioxide molecules are needed to result in 6 carbon atoms, 12 hydrogen atoms, and six oxygen atoms. I've found that fourth- and fifth-grade children find this to be a fun and not particularly complex puzzle, as long as they know that it's okay to have "leftovers." They quickly realize that they need six carbon dioxide molecules, since that's the only source of carbon, and six water molecules to provide the 12 hydrogen atoms. So if they bring together 6 of each molecule using the linker cube representation and then count out how many of each element results, they end up with

◆ six carbon atoms
◆ twelve hydrogen atoms
◆ eighteen oxygen atoms

That's too much oxygen! To figure out what happens to it, we need to know that oxygen travels in pairs, so an oxygen molecule is written as O_2 and looks like this in our cube representation.

FIGURE 4.5

So in the chemical reaction, six oxygen molecules (pairs of oxygen atoms) form in addition to each glucose molecule, to account for the remaining oxygen. When students are actively manipulating the linker cubes they are able to figure out that the "ingredients" needed for each glucose molecule result in leftover oxygen. Energy is stored in the carbon bonds of the glucose molecule, so this forms the energy source that the plant can use to build new structures. The excess oxygen is released back into the air through the stomata.

Another chemical process, called respiration, allows the plant to access the energy stored in glucose. Chemically, it actually looks like the exact reverse of photosynthesis in terms of inputs and outputs. This is not an accurate description of the two processes, but again, for now we're focusing just on the movement of matter and energy through a system. Respiration takes place inside the mitochondria of the cell, rather than in the chloroplasts. When glucose is broken down, the energy stored in its bonds is released (in the form of something called adenosine triphosphate or ATP). When the bonds of the glucose molecules are broken, the atoms re-form into water and carbon dioxide. So we can represent cellular respiration like this:

$C_6H_{12}O_6 + 6O_2 \rightarrow 6CO_2 + 6H_2O$ with energy being released in the breaking down of the glucose

When students learn about these paired processes (more typically at the high school or college level, as this level of detail is beyond what we generally teach elementary school students), they are often puzzled. If plants release oxygen during photosynthesis but then use oxygen and release carbon dioxide during respiration, wouldn't these two processes basically cancel each other out? A simplified answer is that the rate of photosynthesis is much greater than the rate of respiration, so plants expel far more oxygen into the air than carbon dioxide.

This also means that quite a lot of carbon that was once in the air is now in the plant, in the form of glucose and other molecules constructed from glucose. Thus plants, and trees in particular, due to their size and structure, are considered to be *carbon sinks*,

meaning a structure that absorbs more atmospheric carbon than it releases.

So trees, as well as all other plants, are engaged in a constant cycle of using sunlight as the energy source to turn carbon dioxide and water into glucose (with oxygen leftovers). Trees use glucose to build more complex molecules that become leaves, branches, trunks, roots. When all or part of the tree dies, such as when deciduous trees shed their leaves in the autumn, the carbon that was stored there breaks down through decomposition and becomes part of the soil. These interrelated cycles, over the hundreds of millions of years that there have been plants on earth, are the primary reason that we have an oxygen-rich atmosphere and carbon-rich soil, and this is part of the global system that kept the carbon cycle in balance.

Classroom Connections: How Plants Store and Use Energy

Learning about the concept of photosynthesis provides a backbone for later study of plant biology, and it helps children understand why plants in general, and trees in particular, are so important to humans and other animals. Early elementary children can explore the "ingredients" of photosynthesis so that they understand the importance of water, air, and sunlight for plants. They love the idea that plants are food factories, able to produce their own food, which is why they don't have mouths! Acting out a simplified version of photosynthesis, as described in the opening of this chapter, is one way to build this understanding. This pairs well with planting seeds and observing their growth. Beans, which have a much faster rate of growth than trees, work well for this. Planting seeds along the edges of clear cups or jars allows children to see the roots extending down into the soil as well as the chlorophyll-filled stems and leaves growing above ground. This helps children understand that plants absorb water and nutrients from their roots but they do not eat soil!

Other direct investigations can help children further understand the inputs and outputs of trees and other plants. For instance, the classic activity of placing celery or white flowers in water dyed with food coloring, and watching the dye make its way upward, helps children understand how water travels from the roots to all parts of the plant (even though this investigation is about water moving just through a stem, as the roots are detached). I prefer the celery version of this activity, since this emphasizes water reaching the leaves, a necessary step for photosynthesis. Alongside this fun but somewhat disconnected activity, children can observe a wilted plant and then add water to the soil and observe what happens to the stems and leaves. My students have had great luck reviving a basil plant over the course of a school day. Doing these in tandem helps children connect the phenomena of water movement to the overall system of plants converting energy from sunlight into energy they use to grow.

One way to help children understand how leaves are the "sunlight collectors" of plants is to cover with foil a few leaves on a healthy plant. After a few days to a few weeks, depending on the plant, the covered leaves will lose their green color and wilt, while those still exposed to the sun will remain green and continue to photosynthesize. This investigation does not work equally well with all plants, so definitely try it first before inviting children to do it.

Another more common investigation is to grow plants in a sunny window and in a classroom closet. If students start with seeds, they will sprout in both places, since the baby plants depend on the food stored in the seed structure until they make their first leaves. But after that initial stage, the plant in the dark will quickly die, since without light it cannot produce energy to keep it growing. As with the investigation of water movement in plants, it is important to connect this back to understanding the overall *system* that allows plants to convert sunlight into chemical energy. Elementary schoolers do not need to learn the chemical equations of photosynthesis, but they can develop conceptual understanding of the process through acting it out and also closely observing how plants interact with their environment.

Why Forests Are Critical to Climate Stability

All plants photosynthesize, thus reducing atmospheric carbon and increasing oxygen. But forests in particular play a pivotal role in global climate, primarily due to their size. Because individual trees are so large, and a huge amount of carbon-containing glucose is needed to construct the trunks, branches, and other structures, they are particularly effective carbon sinks. In general, larger trees photosynthesize more than smaller trees, and so old growth forests capture far more atmospheric carbon than areas with newer trees. This is not what scientists have always thought, since as trees age, their rate of growth slows. However, recent research indicates that this slower rate of growth is more than offset by the amount of photosynthesis needed to just keep a massive tree alive.[4] Imagine two different households. In one, there is a parent and a teenage athlete. The teenager eats a lot! So the family needs to buy groceries at a rapid pace to keep their family healthy. Now consider a second household comprised of eight older adults. None of them eat as much as the teenager, but because their household is so much larger, they likely consume quite a bit more food overall. So when my students in the opening story were hugging trees to determine their size, they were unknowingly searching for the most effective carbon-capturers in the forest.

Single trees capture large amounts of carbon, and so whole forests have a huge impact on the global carbon cycle. As we noted in the beginning of the chapter, forests cover about 30% of the land mass of the earth. In the past two decades, scientists estimate that forests around the world have pulled approximately

2 billion metric tons of carbon from the atmosphere every year.[5] That amounts to about a quarter of the carbon emitted by human activity.[6] While it would be impossible to fully offset current human greenhouse gas emissions through increased forest-ation, preserving forests as biodiverse ecosystems and carbon sinks is an essential part of bringing earth's systems back into balance.

Unfortunately, human activity over thousands of years has cleared about half of the forests that once covered the earth, and this pace has accelerated in modern times, as the need for land as well as the resources that trees provide has increased. Despite their natural role as carbon sinks, deforestation in some areas, particularly tropical rainforests, is causing some previously forested areas to become net contributors to atmospheric carbon. While rates of global deforestation have slowed in recent years, it is essential that we continue to work on policies that support communities that live in and depend on forests, allowing them to engage in economically sustaining activities while also pre-serving forest ecosystems. We'll look at some of these efforts as stories of hope later in the chapter.

Classroom Connections: Neighborhood Tree Survey

Maintaining forest ecosystems can lessen the impacts of climate change, and trees are a key component of healthy human environments as well. Increasingly, communi-ties are realizing that the presence or absence of trees is a social justice issue. As we'll discuss more in the chapter on cities, urban trees can reduce heat in the surrounding area, mitigate damage from storm runoff, and reduce erosion, in addition to making the air healthier and sequestering carbon. And yet many communities have little to no tree cover. One study of urban areas throughout the US found that low-income census blocks had on average 15% less tree cover than high income blocks, and some cities had a

30% gap. They also found that summer temperatures were consistently higher in neighborhoods with less tree cover.[7] Students of any age can investigate the number and location of trees in their community and use their knowledge to work toward more tree-rich environments.

If your school is near a forested area, students can conduct a survey of the trees in a specific area of the forest, observing the size of their trunk and their canopy cover, the living things in and around them, and other signs of forest health. They can consider ways to help visitors enjoy the forest without causing harm, and they may be able to help rangers or other caretakers do simple tasks such as clearing paths or cleaning benches, which encourage people to stay in designated areas while enjoying the trees.

For those of us who teach in more urban areas, tree inventories can focus on the neighborhood around the school. For young children, taking a tree walk on a sunny day helps them notice where trees are and to physically feel the difference in heat near a tree versus in unshaded areas. They can search for places that might be good for planting new trees. And they can learn about efforts to plant more trees in their communities and perhaps write thank you notes to the community organizations that lead this work.

Older elementary students can engage in a more formal inventory over a larger area, perhaps several blocks around the school. I've sometimes asked my students to count the trees around us when we take field trips to other parts of our region and to notice other features of the human and naturally occurring landscape. Students can integrate math skills through measurements and data analysis, considering questions such as:

◆ How many trees per block are there in our neighborhood (and in others)?
◆ What is the ratio of native to non-native trees?
◆ How much of the sidewalk is shaded by tree cover?

◆ What is the temperature difference in the shade of a tree and in unshaded areas?

Students can also observe photos of neighborhoods where tree planting has changed the streetscape. There are local tree-planting projects in nearly every community, and they are often eager to connect with young people. Based on data that students collect as well as information from these organizations, they may want to contact local government leaders to ask that more street trees be planted near their school or in the neighborhoods that they've identified as having too little tree cover. Municipal governments often have programs that provide street trees on request if residents are willing to engage in the initial upkeep. This is a meaningful project for a class of upper elementary students to take on, as they can see the results of their efforts each day as they come to and from school.

Trees as Part of an Interconnected Ecosystem

Within the complex ecosystems of forests, trees play a critical role in sustaining life and in sequestering carbon, but so do many other forest organisms. Scientists estimate that forests account for up to 80% of the diversity of terrestrial plants and animals on earth. While such an estimate is imprecise, given our incomplete knowledge of earth's biodiversity, there is no doubt that forests support an enormous number of species, and that this biodiversity is critical for the earth's health.[8]

Energy cycles do not end with trees and other plants converting the sun's light energy into glucose for their growth. When an animal eats a plant part such as a piece of fruit, the animal's body breaks down the stored energy and uses it for its own growth and development. If another animal eats the fruit eater, then the energy that entered the earth's system as sunlight is now passed on to this third organism. When animals and plants die, their bodies decompose, breaking down into their component parts. Some of the carbon from dead organisms is released back into the atmosphere during decomposition, but in forest ecosystems, much of the decomposing matter becomes part of the soil, so the carbon that was once in the body of the animal or structure of the plant becomes part of the geosphere.

When areas are deforested, we lose the carbon-holding, oxygen-emitting power of the trees that are removed, but the impact is greater than that. The downed trees may be burned, thus releasing their stored energy back into the atmosphere in the form of CO_2. Wildfires, which have become larger and more frequent due to climate instability, also release massive amounts of CO_2. In both cases, since the soil is no longer held in place by trees, it is more likely to erode via wind and water, and thus the carbon stored there, as well as other greenhouse gases, are displaced into waterways as well as into the air.[9]

In healthy forests, a wide range of forest-dwelling animals including some insects, birds, and bats, act as pollinators, allowing trees and other forest plants to reproduce. Animals are also often the mechanism of seed dispersal, spreading seeds away from the parent plant and thus making it more likely that the next generation of the plant will have adequate resources to survive and thrive. So while animals themselves don't serve as carbon sinks in the way that plants do, they are an important part of the overall system of healthy forests.

One important part of forest systems that gets far less attention than easily visible plants and animals is the fungus, including mushrooms, that grows primarily on and under the ground. The network of fungi in forests is called *mycelium*, and the scale of mycelium networks is vast, far greater than the mushrooms we may see scattered throughout a forest. Mycelium networks can extend as much as 300 feet underground! Fungi survive by breaking down decomposing matter to obtain energy, and that helps to sequester carbon-containing materials back into the soil. The process of mycelium breaking down decaying matter also provides nutrients that help the forest plants thrive.[10] Many scientists think that maintaining and expanding mycelium networks has the potential to increase carbon capture and reduce the impacts of climate change.

While planting more trees may seem like the most obvious and visible solution to deforestation, understanding forests as complex systems of plants, animals, mycelium, soil, air, and other material helps us see that conservation and restoration requires a systems-based approach. Tree-planting efforts are a great place to start, and so too is learning more about the complex and fascinating systems of which they are just one of the more visible parts.

Classroom Connections: On and under the Forest Floor

Elementary students often learn about plant growth and development, but they less often explore mushrooms and mycelium networks. Many adults are completely unaware of the vast, underground network of fungi that is critical to forest health. Children love learning about things that are hidden, and a study of soil and what lives in it can help build understanding of forests as systems. If you have access to a forest, then searching for mushrooms and other "hidden" parts of the forest can be a great start to understanding forests as systems. This type of investigation requires quite a bit of adult support, especially if students are not used to being in a forest environment, since I would recommend against having children directly handle forest mushrooms or any creatures they may find while turning over logs and searching in leaf litter. Many mushrooms are poisonous to humans and so are best observed out of touching distance. I don't like to limit children's freedom to touch, handle, and explore, but in this case, without expert assistance, the risks outweigh the benefits, so I recommend "observation only" activities such as sketching their findings in science notebooks.

A related classroom-based activity uses samples of soil from a forested area to support children's understanding of forest systems. The teacher can first show a container of the soil to the whole class and ask what they notice. Then small groups can observe samples using spoons, tweezers, and hand lenses, looking for details they may not have seen at first. I like to provide these samples in plastic shoeboxes since there are often insects and other small creatures hiding in the soil! While they aren't likely to see identifiable fungi unless you've intentionally included it, they will see that forest soil is not "just dirt" but is instead a rich combination of materials that connect to the other living and non-living things in the forest environment. Children can use books and their own prior knowledge to consider how each observed item is connected to other parts of the forest. See the end of this chapter for some recommendations of books that explore forests as systems.

Growing mushrooms outside of their natural environment can be challenging but is a great way to learn about these important forest decomposers. Ready-made mushroom growing kits result in a small scale, non-toxic mushroom harvest, and these are highly instructive if teachers have the budget to purchase them. Otherwise, this may be a topic best explored through books and videos. The organization Fantastic Fungi has produced a wide-ranging film that explores mushrooms' roles in ecosystems as well as the ways humans benefit from mushrooms. It is long, and there are parts that you may not wish to show children (I would avoid showing the time lapse decomposition of a mouse in an elementary classroom!), but the beautiful footage allows children to experience the huge variety of mushrooms and other fungi and helps them consider the role of fungi in forests.[11]

Forest Champions: Stories of Hope-Filled Action

While global deforestation continues to be a great concern around the world, there are also many reasons for hope that humans can preserve and bring back the forests that we have often failed to fully value. Reforestation efforts are tangible, visible projects that are highly accessible to children. Below are a few stories of forest preservation and growth around the world. I encourage you to also find stories in your own community. This will help children (and ourselves) understand that global efforts are something that we can be a part of no matter where we live.

There are several children's books about the life and legacy of Wangari Maathai, a scientist, politician, and activist who founded Kenya's Green Belt Movement (see end of chapter for recommended books). Under her leadership, this movement sought to re-forest vast areas of Kenya as a means of restoring the environment and improving women's quality of life. Through this movement, women in Kenya planted more than 30 million trees. In planting trees, women in Kenyan villages were able to improve the soil, which had been eroding rapidly without trees to hold it in place, thus improving their ability to farm. They harvested and sold fruit from the trees. By focusing on both reforestation and development of women's economic strength, the Green Belt movement shows that just climate action has economic and social value. While Wangari Maathi's efforts became international in scope (she won the Nobel Peace Prize in 2004), her story is also accessible to children, who can easily picture small groups of people listening to Wangari and then growing seedlings and planting trees to improve their communities. Learning about her work and legacy can be a starting point for children engaging with (re-) forestation efforts in their own community.

Felix Finkbeiner was a fourth grader growing up near Munich, Germany, when he learned about Wangari Maathi's work. His research for a school project led to a presentation in which he proposed planting a million trees in Germany. Local media amplified the story, and his challenge to schoolmates grew into an international effort.[12] While still a young teen-ager, he launched the organization Plant-for-the-Planet, which supports reforestation efforts around the world, with a goal of planting a trillion trees. Although Finkbeiner is now an adult, the organization he founded continues to amplify the voices and efforts of youth as climate ambassadors.[13] While most fourth-grade projects don't lead to creation of an international organization, his story shows the power of young people acting on what they know and believe and of communities working together to improve our planet.

While a charismatic individual often serves as the face and voice of initiatives, reforestation projects always involves groups of people working together toward change, usually in the place they live and depend upon for livelihood and enjoyment. Telling the stories of visionary leaders is inspiring and important, and so too is considering the often many people who collectively bring the vision to life. Learning about communities and people who enact reforestation projects builds connection to the world beyond our own communities and develops the idea that while systemic solutions are needed, people working together can help bring about these changes.

A study of forests, and particularly trees, is well suited to **direct actions** that are accessible to children of all ages. Urban, suburban, and rural communities throughout the country have ongoing tree-planting efforts that range from planting street trees to increase shade and improve air quality, to planting trees as erosion control near waterways, to re-foresting areas damaged by wildfire or human use. Many of these efforts welcome children playing a role, for instance in removing invasive species to make room for native trees.

Sometimes the school grounds can be the site of efforts as well. Neelam Patil, a science teacher in Berkeley, California, helped her students start an afforestation project that expanded to multiple elementary schools in the district. They learned about an approach to establishing dense, biodiverse "pocket forests" that grow quickly and absorb significantly more carbon than typical monoculture forest projects. Patil consulted with members of local indigenous groups to select appropriate, native tree species for these pocket forests. Students raised money for the saplings, did the planting, and are now able to monitor and enjoy these areas.[14] Even if a whole forest is not realistic at your school site, finding ways for children to care for trees and advocate for more forested areas in their community helps build voice and power toward positive change.

Justice-Based Climate Science Unit Example:
What is a tree?
Aligned with NGSS for Grade 1

Guiding Question: What can we learn about a tree by observing it closely? How does our tree change and stay the same?		
Focal Disciplinary Core Ideas All organisms have external parts. Plants have different parts (roots, stems, leaves, flowers, fruits) that help them survive and grow. (LS1.A) Individuals of the same kind of plant or animal are recognizable as similar but can also vary in many ways. (LS3.B)	**Focal Science and Engineering Practice** Asking questions and defining problems	**Focal Cross-Cutting Concepts** Structure and Function Patterns
Engage with Concept and Community	**Choose a Tree.** This exploration ideally takes place throughout the entire school year, so you'll want to choose a tree that is easily accessible. A tree on the school grounds works well, or one at a nearby park. Street trees are great to observe in relation to human communities, but it can be challenging to gather a whole class around the tree safely for long enough to observe and draw. And of course, if your school is near a forested area, this is ideal, but I encourage you to select a single tree in the forest for ongoing close observation. **Four Senses Observations.** Choose a time when it will be quiet in the area around the tree so that children will be able to engage many senses in observing it. You might want to lead a guided four senses observation as a starting point (I recommend not using the sense of taste!) Ask children to use their eyes to look at the colors, shapes, shadows, and structures of the tree. Listen for sounds like leaves rustling or a squirrel climbing the tree. They can carefully touch the trunk, branches, and leaves, perhaps comparing the feel of a leaf still on the tree to one that has dried and fallen off. Finally, smell the air around the tree. Is it different than the air further away? Do the leaves or other structures on the tree have a scent? Once children have had time to engage their senses, they can spend some time drawing their tree. Depending on the age and prior experiences of your students, you may want to provide some guidance, but in general encouraging them to take their time and draw what they see works well. If possible, provide colored pencils as well. Ideally, repeat this experience monthly over the course of the school year. As you do, encourage children to search for evidence of change and also to notice what has stayed the same.		

(Continued)

(Continued)

<table>
<tr>
<td rowspan="5">Explore
ideas grounded in place</td>
<td>

Collect Topics for Inquiry Based on Observations and Wonderings. As children return to the tree throughout the school year, consider using the questions that emerge as the basis for content explorations. For instance:

- Is the tree growing? Children can measure around the tree each month. They will likely find little to know change. However, they will likely see some growth of branches as well as more rapid growth of leaves (and death of leaves if the tree is deciduous). Books about trees can help them understand how trees grow over time.
- Why do the leaves die? This question is a great starting point for discussing how leaves are the energy "factories" of trees. Consider acting out a simplified photosynthesis process. In deciduous trees, in winter when there is less light, trees go through a period of dormancy, in a sense resting until spring comes and leaves again convert sunlight into energy for the tree to grow.
- How old is the tree? There may be elders in the community who remember when the tree was planted, or it may be older than any humans. There are ways to estimate the age of a tree by multiplying its diameter by a growth factor specific to the tree species, so if your students are interested in this, you can help them with the math. Examining tree rings from no longer living trees is fascinating, as it reveals not only the age but also environmental conditions in which the tree grew. Many nature centers and parks have tree cores (a cylindrical slice through the trunk) that children can examine, and parents in your community who work with wood may also be able to provide these.
- How are new trees made? If your tree produces seed pods at any point during the year, children will be excited to consider if baby trees will grow. You might want to bring in seed structures from different trees (ex/ acorns, pinecones, "helicopter" seeds from a maple) so children can examine the ways in which seeds are adapted to disperse/ move away from the parent tree. Seeds disperse so that the new plant has the resources it needs, including light, which is blocked directly under the canopy of the parent plant. Children can test the seeds for whether they move by wind, by rolling, by being carried by animals (on fur or through digestion) or by floating on water.

</td>
</tr>
</table>

(Continued)

	Who lives in the tree? Learning about the birds, mammals, and insects that depend on the tree for food and shelter can lead to learning about interdependence in terms of food webs or, more broadly, ecosystems. If children are able to directly observe animals in relation to the tree, this can be a starting point to considering the role of the tree in the local ecosystem. If there are few animals present, children might consider if there are things they could do to improve the habitat around the tree to make it more conducive to both plant and animal life. I recommend choosing one "wondering" each month to explore via text, direct exploration, and learning from local experts.
Share and learn from community	Consider what experts and elders in your community could share their knowledge and respond to children's questions and ideas. Local park rangers and naturalists may have educational programs and materials already developed, and they are often highly knowledgeable about local tree species and issues of forest health. Indigenous groups in your community may have leaders who are willing to share stories of the land, flora, and fauna from the perspectives and traditions of their people. Some indigenous groups are working to bring their knowledge of land and forest management to issues such as harm reduction from wildfires. If you are in an area impacted by this, learning about these efforts can help children build their knowledge and also reduce fear and worry.
Design hope-filled actions	Help your students consider how they might use what they have learned about their tree to help their community. This might look simply like caring for the area around the tree, keeping it clear of litter and perhaps planting native plants that will improve the habitat. They might want to learn about and help with tree-planting or tree care efforts to increase tree cover in your city or in a nearby natural area. They might even be able to advocate for increasing forested areas in local parks that are primarily grass fields, an environment that does not support diverse ecosystems.
Reflect and synthesize systems	At the end of the year, children will have a year's worth of drawings, observation, and knowledge about this living member of their community. They may have poetry, narrative writing, and other responses to what they have learned over time. Consider hosting a gallery viewing for family and community members, during which children can share what they have learned and their ideas for hope-filled action.

Recommended Children's Books

Bernard, R. (2001). *A tree for all seasons*. National Geographic Kids.

Butterworth, C. (2018). *The thing I love about trees* (C. Voake, illus.). Candlewick Press.

Campbell, N. (2021). *Stand like a cedar* (V. Carrielynn, illus.). Highwater Press.

Chin, J. (2009). *Redwoods*. Roaring Brook Press.

Curtis, A. (2020). *A forest in the city* (P. Pratt, illus.). Groundwood Books.

Garton Scanlon, L. (2018). *Kate, Who tamed the wind* (L. White, illus.). Schwartz & Wade Books.

Gholz, S. (2019). *The boy who grew a forest: The true story of Jadav Payeng* (K. Harren, illus.). Sleeping Bear Press.

Kelley, T. (2022). *Listen to the language of trees: A story of how forests communicate* (M. Hermansson, illus.). Sourcebooks.

Nivola, C. A. (2008). *Planting the trees of Kenya: The story of Wangari Maathi*. Farrar, Strauss, and Giroux.

Pallotta, J. (2010). *Who will plant a tree?* (T. Leonard, illus.). Sleeping Bear Press.

Prevot, F. (2017). *Wangari Maathi: The woman who planted millions of trees* (A. Fronty, illus.). Charlesbridge.

Notes

1 Hancock, L. (2019, March 21). *What's a boreal forest? And the three other types of forests around the world*. World Wildlife Fund. https://www.worldwildlife.org/stories/what-s-a-boreal-forest-and-the-three-other-types-of-forests-around-the-world

2 Education World (2010, March 20). *How does your tree measure up?* https://www.educationworld.com/a_lesson/03/lp309-01.shtml

3 National Geographic Education (2022, July 15). *Chlorophyl*. https://education.nationalgeographic.org/resource/chlorophyll

4 Bourg-Meyer, V. (2015, July 2). *Carbon capture: Tree size matters*. Yale Environment Review. https://environment-review.yale.edu/carbon-capture-tree-size-matters-0

5 MIT Climate Portal. (2021, October 7). *Forests and climate change*. https://climate.mit.edu/explainers/forests-and-climate-change

6 Government of Canada. (2022, May 31). *Forest carbon*. https://www.nrcan.gc.ca/climate-change-adapting-impacts-and-reducing-emissions/climate-change-impacts-forests/forest-carbon/13085

7 McDonald, R. I., Biswas, T., Sachar, C., Housman, I., Boucher, T. M., Balk, D., … & Leyk, S. (2021). The tree cover and temperature disparity in US urbanized areas: Quantifying the association with income across 5,723 communities. *PloS one*, *16*(4), https://doi.org/10.1371/journal.pone.0249715.

8 Food and Agriculture Administration of the United Nations. (2020). *The state of the world's forests*. https://www.fao.org/state-of-forests/en/

9 Welch, C. (2021, March 11). First study of all amazon greenhouse gases suggests the damaged forest is now worsening climate change. *National Geographic*. https://www.nationalgeographic.com/environment/article/amazon-rainforest-now-appears-to-be-contributing-to-climate-change

10 BBC StoryWorks (2023). *Mycelium matters: How mushrooms can address climate change*. https://www.bbc.com/storyworks/climate-academy/mycelium-matters

11 Fantastic Fungi. (2023, March 9). *Fungi field trip*. https://fantasticfungi.com/the-mush-room/fungi-field-trip/

12 Parker, L. (2017, March 7). Teenager is on track to plant a trillion trees. *National Geographic*. https://www.nationalgeographic.com/science/article/felix-finkbeiner-plant-for-the-planet-one-trillion-trees

13 Plant for the planet. https://www.plant-for-the-planet.org/

14 Markovitch, A. (2022, December 8). Berkeley school's "pocket forests" are taking root. *Berkeleyside*. https://www.berkeleyside.org/2022/12/08/miyawaki-pocket-forests-berkeley-unified-school-district

5

The Ocean as a Global System

Grown Up Science	Hope-Filled Classroom Connections
• How is the ocean part of a global system? • How are greenhouse gas emissions impacting ocean systems? • How do ocean and weather patterns interact? • How are shoreline communities uniquely vulnerable to climate instability, and how can we work to protect them? • What nature-based responses can mitigate the impacts of climate change on oceans?	• Design living shorelines to protect aquatic habitats and human communities • Discover how we are connected to the ocean, no matter where we live • Explore how living things are adapted to specific environmental conditions: the case of brine shrimp • Learn about ocean heroes: oysters and kelp as natural climate change fighters

"Woah, it didn't look like it had mattered that much!" Jasmine looked at our data chart of how different models of artificial ocean reefs had decreased the size of waves in our testing tank. One of the models, a "castle" design, had reduced the height of the waves by almost half. Because we were measuring small,

DOI: 10.4324/9781003393535-5

fake waves inside a large plastic bin, the difference between 1.5 cm and 3 cm had not seemed significant when the children first observed it. But looking at the compiled data and thinking about what that sort of decrease might look like for real ocean waves, they were amazed. We looked at photos of artificial oyster reefs installed in different parts of the San Francisco Bay, including one installment just a few miles from our town. Children had all sorts of ideas of how they might change and improve their models to reduce wave energy even more.

At our summer science camp, these children, in grades 3–6, had spent a week learning about waves and how they impacted the shoreline in our area. We watched videos of waves and discussed how they formed, how they were impacted by weather, and what how they impacted different types of shorelines. We constructed "wave tanks" out of large plastic bins and used re-purposed cafeteria trays to make waves of different sizes. I then introduced the idea of communities needing to adapt to bigger storms by protecting themselves from the impact of big waves. The children brainstormed ways they might do this and, unsurprisingly, came up with the idea of a seawall, which is in fact a common response to protecting shore-side communities. One student mentioned that if the water managed to get over the wall, it would get stuck on the other side and might make flooding worse. This provided an opening to introduce the idea of living shorelines and permeable structures that reduce wave energy and allow water to move freely back into the ocean or bay.

Over the next several days, we zoomed in on a study of oysters, an important bay and near-shore species that has seen major population decreases on both US coasts in recent years due to decreased habitat, overfishing, and changing environmental conditions. We learned about local scientists designing human-made reef structures to restore oyster-based habitats and serve as protection for the shoreline. The children made models of the different reef structures scientists had tried and then, based on the results of testing in our wave tanks, they worked to improve their models. At the end of the week, they expanded their model "wave reducers" to include other aspects of living shorelines,

FIGURE 5.1 One group's design for an oyster reef structure.

including their revised reef structures as well as model wetlands. We ended this study with a "dream and draw" activity, where small groups of children imagined a shoreline that provided a safe and healthy habitat for animals, plants, and humans. Their drawings included evidence of many things we had learned during the week as well as glimpses into broader hopes and ideas: children playing along the shore, elevated houses with giant playrooms, oyster reefs and marsh areas next to sandy beaches, and in one drawing a very happy whale passing by in the deeper ocean.

Building Hope-Filled Connections to Ocean Environments

Oceans play a critical role in the health of the climate. They are also fascinating to many children: creatures that light up, underground "forests," massive waves, and countless other features hold attention and provide opportunities to build both understanding and connection to these environments even for children who do not live near them. For the 40% of the US population who live in coastal or near-coastal communities,[1] opportunities for connection and knowledge-building are both urgent and compelling, as communities increasingly experience the impact of unstable weather patterns, ocean acidification, and other results of climate change. And as we'll explore in this chapter, all of us are deeply connected to ocean systems, no matter how far away we live from a coastline.

As with all of the topics in this book, my recommendation is that we consider carefully how we approach ocean-related climate change issues with children. The principles of **understanding systems, place-based learning, and hope-filled action** are particularly important in approaching the study of something as big as the world's oceans, where so much is happening that we as non-experts may feel powerless to address. Adults need to fully understand the causes and impacts of ocean acidification in order to advocate for changes to slow and ultimately reverse this process. Children can start building understanding not by jumping straight to coral reef bleaching (unless they actually live near and experience this process first-hand), but rather by learning the amazing ways in which ocean-dwelling organisms are adapted to specific environments and how even small changes to these environments can cause them harm. They can build their knowledge of how ocean ecosystems and humans are interconnected, and they can learn about people who are working to maintain and restore healthy ocean ecosystems. As adults, we need to develop solid understanding of what is causing sea level rise and what local and global actions are needed to slow this human-caused change to our oceans and land. Children can explore the complex and fascinating environments where water and land meet and develop understanding and appreciation of the biodiversity and geological importance of shoreline environments. They can design and test models of nature-based methods to protect shorelines with a focus on keeping communities safe and healthy. As we support children to build knowledge, connection to natural systems, and a sense of agency, we help to develop the next generation of scientists, policy makers, writers, and community leaders who appreciate the importance of oceans and understand how oceans and human life are inextricable.

The Ocean as Part of Earth's Interconnected Systems

When I see photographs of deep-sea creatures, oceans can seem as mysterious as outer space. On the other hand, for more than 20 years I have lived less than a mile from the San Francisco Bay, and the rocky shores and tidepool creatures that once seemed so unusual to someone raised in the mountains of Virginia now seem like old friends. If your community is near an ocean, you and your students may have lots of experience observing waves, searching for shells, and spotting dolphins swimming close to shore. Or perhaps you live far inland and few if any of your students have seen an ocean up close. Either way, we are all connected to the ocean.

If you haven't looked at a globe recently, take a few minutes to observe how much of one is covered in blue. The world's ocean covers about ¾ of the earth's surface, and it is critical to global climate systems and to all life on earth. While we traditionally learn about four or five separate oceans, your consultation with a globe will confirm that there is really just one giant body of water, as all of the named oceans are connected to one another. Helping students understand the idea of "one ocean" is a step to understanding the earth as a set of interconnected and inseparable systems.

While the ocean is vast, it is not at all uniform. What most globes don't show is the amazing topography present under the ocean water. There are mountains, valleys, and plains similar to what is found on the dry land. This underwater landscape creates widely varied conditions for life, as does the composition and behavior of the seawater itself. Differences in temperature, salinity, and dissolved matter create sections of ocean water that are layered on top of one another. Wind patterns move warmer waters from the surface and near-shore areas farther from the land, and this water is replaced by colder, nutrient-rich water from deeper in the ocean. Along the coastlines, this process of water exchange provides nutrients to support a rich and varied near-shore ecosystem. Further out to sea, a vast array of organisms are adapted to the varied ocean conditions in almost unimaginable ways: sharks that never stop swimming so as not to sink to the sea floor, worms with special appendages to bring in as much oxygen as possible in places where there is hardly any, fish with special organs that glow to provide light in the deep-ocean and underwater caves, and so many more. Organisms have adapted over millions of years to specific temperatures, salinity, light level, and available food sources.

Ocean water, as well as the water in lakes, streams, and bays, is not pure H_2O. It is a complex solution to which the animals and plants living in that water are adapted. For instance, all water that supports life contains dissolved oxygen and carbon dioxide, which are necessary for the survival of water-based plants and animals. Oceans also contain dissolved salt, and ocean-dwelling plants and animals are generally adapted to survive within a narrow salinity range. **Estuaries**, the places where fresh and

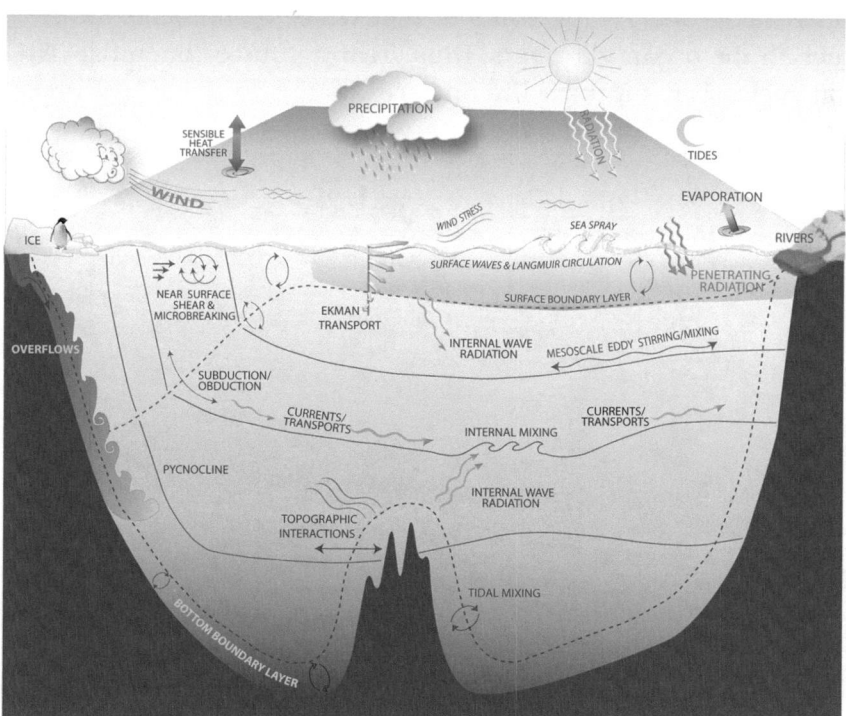

FIGURE 5.2 The ocean is made of many layers that mix and interact with each other and with land and atmosphere.

saltwater meet, have lower salinity levels and also have a greater range of salinity, since fluctuating conditions such as increased river flow from spring snow melt change the balance between fresh and saltwater flowing in.

Estuaries are clear markers of how ocean, freshwater, and land environments interact, but even communities nowhere near the literal meeting place of land and water are greatly impacted by the ocean because of its roles in the water cycle, temperature regulation, and gas exchange. Since most of the world's water is in the ocean, it is the primary source of the water that evaporates into the atmosphere and later falls as rain on both land and water. Warm temperatures cause large amounts of water to evaporate from the surface of the ocean (since only the actual water evaporates, salt and other dissolved materials are left behind). Colder air then pulls some clouds inland, and their precipitation falls on land as well as freshwater lakes and rivers. Some of that water then makes its way back to the ocean via rivers, thus creating one of the most critical cycles that supports life on earth.

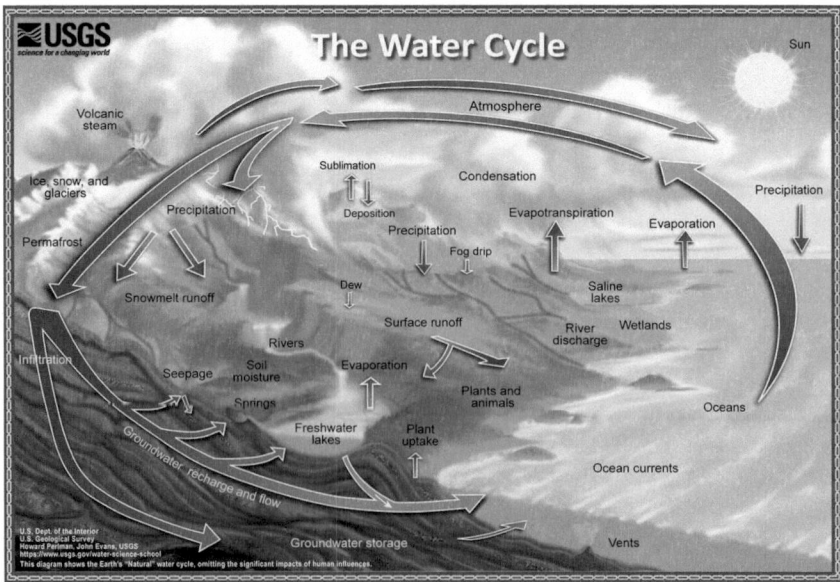

FIGURE 5.3 The ocean is a critical part of the water cycle that moves water to all parts of the earth. USGS Water Science School (2019, October 16). *The natural water cycle.* https://www.usgs.gov/media/images/natural-water-cycle-jpg

The term *watershed* refers to the area of land where water from rain and snowmelt are channeled along a particular path of rivers and streams to the ocean. We all live in a watershed that connects to some part of the ocean. The classroom connection section that follows engages children in figuring out how the place they live is connected to the ocean through water.

Oceans also impact all life on earth due to their role in regulating global temperature. Water has a higher *heat capacity* than air. Heat capacity refers to the amount of energy needed to change a type of matter's temperature. The heat capacity of the ocean is 1,000 times greater than air, which means that oceans can absorb and hold much more heat energy than the atmosphere. The high heat capacity of ocean water helps moderate the earth's climate, as it absorbs much of the sun's energy. However, as we'll discuss later in this chapter, this also means that as more heat is trapped in the atmosphere, much of this excess energy is absorbed by the ocean, causing water temperatures to rise, which impacts ocean ecosystems as well as weather patterns.

We are all connected to the ocean as a source of food and means of transporting the materials that support modern society. 8% of protein consumed by people worldwide comes directly from the ocean, in the form of fish and shellfish.[2] Oceans also serve as the world's primary shipping highway. Approximately 90% of imported and exported products are transported via cargo ships. So no matter where we live, the weather we experience, the water available through our taps, and the goods that support our lives connect us to the ocean.

Classroom Connections: How Are We Connected to the Ocean?

I'm always surprised at how much my students love maps. Maps are models that allow us to understand certain features, for example distance, elevation, or topography, without being full replicas of a place. Map explorations that can help children understand how they and others are connected to the ocean, no matter where they live.

One way to help children understand how very much of the earth is covered in ocean is to play with a beach ball globe (these are inexpensive and lots of fun for a variety of classroom activities). Standing or sitting in a circle, children toss the ball to each other and with each toss determine whether the catcher's right thumb landed on land or ocean. For younger children, placing a small sticker on each student's right thumbnail helps a lot! A simple tally chart quickly reveals their thumb is much more likely to be touching ocean than land. Older children can calculate the percentages, and this can also a great way to show that larger sample sizes tend to yield more accurate results: the first four or five students who catch the ball might get results that are entirely land or water since this is a random chance activity, but tossing the ball 100 or more times will likely give results closer to the 75% water and 25% land that is the approximate ratio of water to dry land on earth.

Closer map exploration can make connections between where your students live and the ocean. For this, you'll need maps that show the geographic features of your region and also show the nearest ocean. While it's possible to do this with printed maps, the Streamer App made by the US Geological Service facilitates this.[3] This interactive map allows students to zoom in on their town or any other location in the US and then click on a waterway to see how it flows downstream and ultimately connects to the ocean. This works well as a whole class activity with elementary schoolers, who will likely need teacher support to locate different places and understand issues of scale. The teacher can project the map and let children take turns finding locations and waterways and predicting their full path before clicking on it to reveal its connection to the ocean.

How Climate Change Is Impacting Oceans

As the impacts of anthropogenic climate change increase, they are causing significant, lasting changes to the world's ocean. In this section we'll discuss some of the largest impacts. I recommend working to understand these impacts yourself to make informed decisions as a teacher and global citizen, but be cautious in how you introduce them to children. Engaging children's fascination with and connection to the ocean from a systems perspective and providing them with hope-filled stories of people working to lessen the causes and impacts of climate change provides a basis for later, more complete understanding of the issues I'll introduce here. The classroom connection ideas throughout the rest of the chapter provide potential starting points, but also consider the specifics of where you live and how you might use connection to your students' specific place in the world to help them understand the power and promise of our oceans.

And remember that adults as well as children can become hopeless when considering the enormity of the challenges that face our planet. Please feel free to read the information below a little at a time, and to skip ahead to the classroom connections and cases of hope-filled change in order to remember our own power to be part of positive, restorative change.

Ocean Acidification

As we discussed above, ocean water contains a mixture of dissolved solids and gases. On the surface of the ocean, where water meets sky, there is a constant exchange of gases, and since the surface area of the ocean is so vast, so too is its ability to absorb atmospheric gases. The ocean is a giant carbon sink, absorbing over 30% of the CO_2 that is released into the atmosphere.[4] The same processes of photosynthesis and respiration that occur in land-based ecosystems also occur in the ocean, so there is constant cycling of oxygen and carbon dioxide through living and non-living parts of the system. However, as with land-based environments, the excess CO_2 being emitted through burning fossil fuels is accumulating at levels far beyond what the system is adapted to support.

In ocean water, absorbed CO_2 goes through a series of chemical changes that result in the water becoming more acidic (a quick reminder from high school chemistry that acidity is measured through the pH scale, and *higher acidity lowers the pH*). Changes to the ocean's pH may seem quite small to those of us who are not chemists, from an average of 8.1–8.2 to an average of 8.0–8.1 since the start of the Industrial Revolution. However, this actually indicates a 26% increase in acidity. Ocean organisms are adapted to specific pH levels, so even small changes impact their survival. And the change is not evenly distributed, with some parts of the ocean experiencing significantly greater changes to pH than others.[5]

The changing pH of ocean water is particularly harmful to the many organisms that make their shells from calcium carbonate, as increased acidity leads to less availability of calcium carbonate in seawater. This impacts shellfish populations that are important food sources for humans. For instance, oysters are particularly vulnerable to acidification at their earliest stages of life, so this results in fewer oysters reaching maturity. Many plankton species also form shells, and declines in plankton populations, a food source for countless other ocean organisms, ripple up the food chain to impact whole ecosystems.[6] The coral that form the basis of reef ecosystems are sensitive to changes in acidity as well as to temperature, which we'll discuss below. The decreased availability of calcium carbonate impacts the ability of coral to add to their skeletons, and this also limits their ability to reproduce.

There are ways to use habitat restoration and nature-based engineering to mitigate changing pH levels on a small scale. For instance, sea grasses such as eelgrass have been shown to increase the pH (reduce acidification) in the water around it as a result of their using CO_2 for photosynthesis. In areas with abundant eelgrass, there is greater availability of calcium carbonate for shell formation.[7] A study in the Chesapeake Bay found that the calcium carbonate produced in areas rich with sea grass then spreads via currents to other areas, thus improving conditions for shellfish over a wider area than just the sea grass bed.[8] Sea grasses also provides a great habitat for egg laying and offers some protection from waves. Likewise, kelp, as we'll discuss in more detail below, can reduce acidification and improve habitat conditions. While plant restoration efforts cannot by themselves address the massive scale of ocean acidification, if combined with global efforts to halt greenhouse gas emissions, they can play a powerful role in protecting our oceans.[9]

Increased Temperature and Decreased Salinity

As we discussed earlier in the chapter, the ocean absorbs far more of the sun's energy than the atmosphere because of water's high heat capacity. This characteristic of water means that as the overall atmospheric temperature increases, much of the heat is transferred into the ocean. In fact, since 1960, scientists estimate that the ocean has absorbed 20 times more heat than the atmosphere. Because of water's high heat capacity, the temperature of the ocean changes quite slowly, but the increase is steady and ongoing, and as with other chemical properties of the ocean, changes that may seem small actually have enormous consequences for ocean ecosystems and for the ocean's role as a regulator of earth's climate.

The increased temperature of the ocean is the primary cause of sea level rise due to a phenomenon called *thermal expansion*. When matter heats up, it expands, and that is exactly what is happening to ocean water as it absorbs more and more heat from the atmosphere. Scientists estimate that up to 75% of sea level rise is due to the thermal expansion of the ocean.

It is actually increased air temperature more than water temperature that is leading to melting of polar ice, but this melting ice impacts both land and water environments. Unlike liquid water, ice and snow primarily reflect light energy from the sun. So as ice and snow melts into water, more energy is absorbed, further increasing the ocean's temperature.

Freshwater from melting ice also decreases the salinity of the ocean. In addition to impacting ocean ecosystems that are adapted to very specific ranges of salt in the water, this has the potential to change patterns of ocean currents, thus impacting weather patterns all over the world. Scientists have not found evidence that this is happening yet, but it is something they are carefully monitoring as the impacts of climate change continue to increase.[5]

On a smaller scale, increased storms and "boom or bust" precipitation patterns make the salinity of estuaries—the places where ocean and freshwater meet—more volatile, which impacts the rich ecosystems that thrive in estuary environments. Estuaries are critical breeding grounds for many marine animals and an important source of food for human communities, so changes to their health have repercussions far beyond their boundaries.

Decreased Oxygen

While levels of CO_2 in the ocean have greatly increased due to human's disruption of the carbon cycle, oxygen levels have declined an average of 2% between 1960 and 2010. This phenomenon, called **deoxygenation**, is caused by climate change as well as other human activities. First, the changing temperature of the ocean impacts oxygen levels, since warm water holds less dissolved oxygen than cooler water, and so as surface water heats up it absorbs less oxygen from the atmosphere. This surface water warming also impacts how layers of ocean water mix and can make oxygen less available to organisms in deep-ocean environments. In addition to climate-related causes, humans have caused deoxygenation through runoff of nutrient-rich fertilizers from farms into waterways. This causes rapid increase in algae growth, which uses up available oxygen in the water as it dies and decomposes. Algal blooms also block light from reaching ocean plants, thus decreasing the oxygen provided by photosynthesis. Loss of oxygen in ocean water makes the environment unsustainable for populations of fish and other marine animals.[10] Reducing fossil fuel use and also eliminating excessive nutrient runoff are essential steps to curbing deoxygenation.

Oceans and Global Weather Patterns

Oceans are critical to the global water cycle as well as to weather patterns. As the ocean warms due to increased greenhouse gases in the atmosphere, these impacts cycle back into the atmosphere in the form of changing weather patterns. Heat increases evaporation. As the atmosphere and ocean warms, water cycle patterns become increasingly unstable and difficult to predict. The instability leads to more extreme weather events, from hurricanes and other massive storms to longer, more intense periods of drought. Wherever you live, you and your students are likely already experiencing the more extreme weather that is the result of climate instability. While individual weather events cannot be singularly attributed to climate change, the overall pattern of increased extreme weather results from the ways in which humans have altered the atmosphere and ocean. Reducing human impact on these global systems is critical.

All of the impacts discussed above require systemic approaches to slow or reverse. Because many of the impacts of climate change on the earth's ocean are beyond young children's ability to directly impact, I recommend introducing children to more accessible systems of interaction that will help them understand how changes to one part leads to changes and impacts on other parts. As they gain more knowledge and connect it to their early explorations, they'll be able to add more details to their understanding of how living and non-living parts of the ocean and atmospheric system interact.

With that in mind, the investigation in the next classroom connection section is not directly about the broad impacts of climate change on the ocean. Rather, it is an exploration of the interactions between environmental conditions and the needs of organisms. This is a far more inviting and developmentally sensitive starting point for young children coming to know and care about saltwater habitats. By inviting children to build skills of observation, investigation, and analysis of a very tiny system, we build their capacity to gradually understand systems as vast as the ocean and all that connects to it.

Classroom Connections: Exploring How Living Things Are Adapted to Their Environment

When you were a child, did you ever convince your parents to let you have a "Sea Monkey" kit? While advertising that suggested we could grow mermaids in a tiny aquarium was misleading, these kits do in fact contain a fascinating creature more commonly known as a brine shrimp. While "brine" means salt water, brine shrimp are native to inland saltwater lakes, not the ocean. However, because they are easy and inexpensive to grow in a classroom environment, they are a great organism to use to explore how living things are adapted to specific salinity conditions. Below are two ideas for engaging children in brine shrimp investigations and connecting this to understanding ocean and estuary environments.

One way to explore this with young children is through a "mystery organism" investigation. In this multi-day lesson, children first consider what they know about living things and then use this to design an investigation. I first provide small groups of children with two magnifying boxes, one containing clover seeds (which will sprout in fresh water), marked "A" and one containing brine shrimp eggs, marked "B." I ask students to observe without opening the boxes, using the prompts "I notice …" and I wonder …" to guide their discussions. I then tell them that both boxes contain living things, but I'm careful not to reveal what they are! I ask, "How do you think we could bring the mystery organisms to life?" Students will come up with many ideas. They often think the clover seeds—which are relatively large and spherical—are eggs, so they may suggest putting them in a warm place, much like a bird sitting on an egg. Some children will infer that one or both of the boxes contain seeds and suggest planting them in soil. Only occasionally have I heard young children come up with the idea of putting them in water, but when suggested they usually readily agree that could work.

After the initial observation and discussion, it's time to set up the conditions to make the organisms grow! I show the class the materials they can use: salt water, fresh water, and soil, all at the front of the room in pourable/scoopable containers, and I provide each group with four small containers. Groups then draw a plan of what they will put in each container. My favorite way to do this is to allow groups to design their first attempt in any way they want, for instance adding soil, salt water, and both types of organisms to one container even though I know this won't lead to anything growing. However, if we don't have time in the curriculum to repeat the investigation so that all groups can successfully grow both organisms, I'll provide only salt and fresh water (no soil at first) and more directly guide the setup, so that each group tests both mystery organisms in both environmental conditions.

Once the test containers are set up, children observe and draw what they see every day or two. Within about three days, depending on temperature, the brine shrimp eggs placed in the salty water will hatch. Shortly thereafter, the clover seeds emerge in containers containing either fresh-water or a soil/freshwater combination.

Once children see the brine shrimp swimming in the salt water, the excitement is uncontainable. They eagerly observe them with magnifiers, and I use a projecting microscope so they can see them using their many tiny appendages

FIGURE 5.4 Brine shrimp as seen under a microscope. © Hans Hillewaert, CC BY-SA 4.0. Hillewaert, H. (2010, February 26). Photo of Artemia salina. CC BY-SA 4.0. https://commons.wikimedia.org/w/index.php?curid=9575707.

to swim. The brine shrimp initially eat the outer parts of the eggs from which they hatched, and then we add tiny amounts of yeast as food.

After plenty of time to observe and enjoy these tiny, swimming creatures, we discuss why it might be that brine shrimp only hatched in the salt water and the clover (less exciting than the swimming shrimp, but continually growing nonetheless) only in fresh water with or without soil. This allows children to consider how important it is to keep different environments healthy for the animals and plants that live there, since they cannot simply move somewhere else, where they would not be adapted to the environmental conditions.

Upper elementary students love and learn from the mystery organism investigation just as much as kindergarteners, and they can also do a more nuanced form of this investigation, focusing only on the brine shrimp and the impact of salinity on hatching. They can directly test the range of salinity in which brine shrimp can live (which is quite a bit saltier than typical ocean water). A device called a salinity refractometer can measure how much salt is in the water, and inexpensive versions are available from aquarium supply stores. This allows collection and analysis of quantitative data in this investigation and, if you live near a saltwater environment, it is also a great tool for field investigations, such as comparing the salinity of a bay in different locations or after different weather events.

Where Water and Land Interact

Human-caused changes to our global system are impacting how water is distributed on the planet through both sea level rise and increased strength and frequency of storms. In this section we'll briefly discuss both of these issues, and in the classroom connection section that follows, we'll focus on hope-filled actions to mitigate the impact of these issues on communities while we work at a national and global scale to reduce the long-term changes to earth's atmosphere that are causing them.

As we discussed earlier, thermal expansion of ocean water due to increasing temperatures is primarily responsible for sea level rise. Sea level rise impacts where high tide reaches on the coastline. High tide flooding has dramatically increased over the past 50 years. Global sea level has risen 8–9 inches since the start of the industrial revolution, with 3.8 of those inches being just since 1993.[11] However, the impact of sea level rise isn't the same along all coastlines in the world. First, wind and ocean currents result in an unevenly heated ocean, so some parts have expanded more rapidly than others. Also, coastal features interact with the change in water volume in different ways. Some parts of the US coastline are being impacted at a much greater rate than the overall increase would suggest due to factors such as erosion and oil and gas drilling.

Rising sea level threatens the infrastructure of coastal cities and towns, potentially putting homes, roads, bridges, and other critical parts of infrastructure under water. Just as critical is the impact of storm surges, which most coastal communities are experiencing with increasing frequency. Storm surges have always been part of bad weather events, with heavy rain and wind causing much higher tides, and decreased paths for water to escape as inland waterways fill and drainage and barrier systems overflow. Damaging storm surges have gotten stronger and more frequent due to increased instability in the atmosphere combined with rising sea levels.

Storm surges erode the shoreline and disrupt sediment on the floor of oceans and bays. This reduces habitat for near-shore organisms. Living shorelines are a nature-based approach that can be highly effective at reducing the impacts of both storm surges and sea level rise, serving as sponges that buffer the impact of water on areas further inland. Re-populating marshes with native plants such as grasses holds the mud in place, lessening erosion. It also reduces the energy of storm surges and filters contaminants from water. In addition, these restored areas provide habitat to many shore-based and near-shore animas to nest, forage, and hide from predators.[12]

Case Study of Oysters: Building Back Economies, Ecosystems, and Coastline Protectors

Oysters, which are found on both coasts of the United States, are critical organisms for maintaining healthy near-shore ecosystems, areas that can serve as a buffer to some of the impacts of climate change. Learning about oysters and about current conservation and restoration efforts helps children and adults understand how important coastal ecosystems are to human communities.

Oysters are "water filtering machines." Oysters eat by pumping ocean/ bay water through their body in order to access nutrients from algae and microscopic plankton. In doing this, they also filter out phosphorous, nitrogen, and other potential toxins, and they are able to use these to build their shells. A single oyster can filter 30 liters of water in an hour.

When oysters die, their shells sink to the bottom of the bay or ocean floor. As they decompose, the carbon from their shells is largely absorbed into the floor rather than being released into the water or atmosphere, and this decomposition also increases the nutrients in the mud of near-shore, marsh, and other wetland habitats. So while they filter out excess nutrients from water-based environments, they also provide needed nutrients for land-based ones.

Healthy oyster reefs also provide powerful shoreline protection. As students discovered in the opening story, oyster reefs can greatly reduce the wave energy that reaches the shore. Re-building healthy oyster populations is often a central part of living shoreline projects. Scientists on both US coasts are studying ways to improve the subtidal habitat to create more supportive conditions for living organisms including oysters and also to protect the shoreline from the impacts of sea level rise and storm surges.

One example of this is the research and restoration project at Giant Marsh, an urban shoreline in the San Francisco Bay Area. The California Coastal Conservancy is working with a wide array of community partners on a project that will inform other shoreline protection projects in the region. They have installed a multi-faceted living shoreline that includes eelgrass beds, native seaweeds, and native marsh plants, in addition to engineered reef structures to promote re-population of oysters and the other organisms that oyster reefs support. In an older oyster reef installation not far from the Giant Marsh project, scientists found that wave energy was reduced by 30–60%.[12]

On the East Coast, the Chesapeake Bay Program is working to restore oyster habitats in ten tributaries of the bay, helping to address the increased nutrient load that has negatively impacted local ecosystems, restore habitat for organisms that thrive in and around oyster reefs, and make the shoreline more resilient to rising seawater. Scientists continually monitor these installations to determine best placement and underlying structure for artificial reef habitats as well as the impacts of these efforts.[13]

Artificial oyster reef structures, often "seeded" with young oysters, have been highly effective at restoring habitat. Designing models of artificial reef structures and testing their impact on reducing wave energy is a highly engaging classroom activity that allows children to understand the power of nature-based engineering and to see themselves as innovators as they envision ways to make the structures even more effective. I'll discuss other possible oyster investigations in the classroom connections section below.

Case Study of Kelp: Hope in the Form of Seaweed

Even if you spend a lot of time near oceans, you may not give much thought to seaweed. However, it is critical to ocean food chains and may prove to be an important part of addressing ocean acidification. Kelp, a large brown seaweed that grows in shallow ocean waters, covers ¼ of the world's coastlines in areas known as kelp forests. These kelp forests are a source of food and shelter for thousands of ocean species. In addition, kelp helps to keep ocean and atmospheric chemistry balanced and supportive of life.

Kelp grows rapidly, which means that its rate of photosynthesis is high. In fact, kelp can absorb 20 times more CO_2 per acre than land forests. In addition to pulling carbon (in the form of carbon dioxide) from the water, it also pulls carbon and nitrogen compounds out of the atmosphere, since kelp grows near the surface of the ocean. This reduces ocean acidification as well as the concentration of greenhouse gases in the atmosphere.[9]

Kelp forests are decreasing worldwide, in part due to warming waters, and also because of overfishing that has caused imbalance in ecosystems. As predators decrease due to overfishing, there has been an enormous increase in populations of sea urchins, which feed on kelp. While overfishing itself is not directly causing climate change, the resultant decrease in kelp means less carbon is being pulled out of the atmosphere. Kelp forests off the Northern California coast have shrunk 95% since 2014 due to repeated heatwaves as well as a marine virus that caused the die-off of a sea star than is the main predator of sea urchins in the region.[14]

Scientists and entrepreneurs are looking for ways to restore kelp forests in ways that also improve economic conditions. For instance, some groups are working to market the kelp-eating sea urchins as a food source, and also promoting the use of urchin shells in the production of cosmetics and fertilizers. Kelp farming is also a promising practice. In this process, kelp is grown in huge tanks and then moved to the ocean when it is larger, increasing chances of survival. A startup in Maine is developing a system to grow large amounts of kelp attached to weights that would, once the kelp is mature and thus heavy, sink to the bottom of the ocean and in effect return the carbon we've removed and displaced through extraction of fossil fuels.[15]

Classroom Connections: Building Understanding of Oysters' Role in Ecosystems

This chapter's opening story and sample unit outline engage children in investigating living shorelines as a way to support ocean ecosystems and protect human communities. Oyster reef structures form an important part of many living shoreline designs, and oysters are fascinating organisms that play an important role in stabilizing near-shore environments beyond their role in reducing wave strength. Understanding how their mode of eating, filter feeding, impacts their environment, is a great entry into understanding interrelated systems.

I have never tried to keep actual oysters in a classroom environment, and I would caution against doing so unless you have access to the expertise and materials needed to keep them healthy outside their natural environment. However, there are a number of other ways that children can learn about oysters' critical role in near-shore ecosystems. One is to explore different materials that filter water, comparing the speed and effectiveness of different materials, and then connect this to how effectively oysters act as natural filters. They can also examine oyster shells to get a sense of their anatomy, including how they attach to other surfaces to form reef structures.

Once children have learned about the structure and function of oysters, they can explore how they are connected to a larger ecosystem. This can be done with photos of plants, animals, and non-living elements that students connect with yarn or arrows to show how each organism depends on other organisms and natural features to survive. Students can also make physical models of healthy shoreline areas highlighting the concept of balanced ecosystems. If you live near oyster habitat, a visit to a place where oysters live is a great beginning or end to this study, although the oysters may not be directly accessible since they attach themselves to rocks and may be too far for the shoreline for safe exploration. Local park rangers, ecologists, and restoration volunteers are often eager to share their work with children, either on site or back in the classroom.

If you live far from the ocean, learning about oysters and how people are working to restore their environment can connect to study of a water-based organism in your area. For instance, while some species of snails are invasive and cause environmental harm, native snail species play a critical role in many water-based environments. While they don't filter water in the same way that oysters do, they eat dead and decaying matter and so help to keep the water clean, and like oysters, they are an important food source for many other organisms. Comparing a local species of snail or other similar organism to an oyster species helps children build connections between varied ecosystems.

The Promise of Ocean-Based Renewable Energy

Spending time on the shore, or venturing into the ocean, reminds us of the vast energy that is in and around the ocean. Waves hurtle forward and crash onto the shore over and over. Wind moves across the water in predictable patterns. The energy of the ocean is inspiring scientists and engineers to envision new ways that humans could access the energy they need without continuing to pour greenhouse gases into the atmosphere. Prevailing ocean winds have long allowed people to power sail boats. Newer innovations are being developed to convert the energy from waves and tidal movements into electricity, a source of renewable energy that could help to replace a reliance on fossil fuels. We'll explore this in more depth in Chapter 8, in which we consider sustainable energy.

Finding Hope-Filled Action in Your Place

If you teach near a coastline, then direct investigation of oceans and bays builds understanding of interrelated systems and deep knowledge of place. If you live further inland, studying the ocean is still fascinating to many children, but you will also want to consider the waterways near you as starting points for first-hand learning. Many of the ideas that we've discussed in this chapter can be brought to life through study of places close to where you live: how we are all connected via the earth's waterways, adaptation of living things to specific ranges of environmental conditions, and nature-based engineering as a way to both restore habitats for local plants and animals and protect human communities from some of the impacts of rising water and unpredictable, unstable weather patterns. And in nearly every community, there are groups of scientists and other community members working together on these issues, and their stories can inspire children to imagine their own role in protecting and restoring waterways as part of our response to climate change. We are all connected to the ocean, no matter how far away from one we are, and building these connections with young learners will help them understand, care, and feel empowered to enact changes to conserve this vast and critical part of the earth's system.

Justice-Based Climate Science Unit Example: Designing Living Shorelines Aligned with NGSS for Grade 2

Guiding Question: How can we use nature and engineering to protect communities from storms and waves? (1-ESS2-1, 2-ESS2-2)		
Focal Disciplinary Core Ideas Wind and water can change the shape of land (ESS2.A) Because there is always more than one possible solution to a problem, it is useful to compare and test designs (ETS1.C)	**Focal Science and Engineering Practices** Analyzing and Interpreting data Constructing explanations and designing solutions	**Focal Cross-Cutting Concept** Stability and change
Note: This unit is more fully explained in an article I wrote with my colleague Sarah Ferner. If you are interested in enacting this in your classroom, I recommend reading the article and referring to the online lesson resources, in addition to the summary below.[16] Children engage in modeling wave action throughout this unit. To accomplish this, Sarah and I constructed model wave tanks out of large, deep (110 quart) plastic bins containing 15 cm of water. Students used plexiglass sheets cut to the width of the bins to create waves by pushing. Cafeteria style trays also work well as wave makers. Rulers taped vertically at both ends of the tank allow students to measure wave height. While completing water-based activities outdoors is ideal, I've also done this indoors many times, just be sure that the bins are deep enough to minimize splashing, and avoid carpeting and electronics!		

(Continued)

(Continued)

Engage with concept and community	**How do waves move?** A great way to begin any unit is to ask children to share what they know. Even in the community where I live, adjacent to the San Francisco Bay and near the Pacific Ocean, children have widely varied levels of direct experience with water and waves, so I like to begin with students watching a video of ocean waves and sharing what they observe and what they wonder. This allows children to share prior knowledge while also providing opportunities for all students to contribute ideas. Next, invite children to practice making waves using model wave tanks (see note above). Challenge them to practice making steady, predictable waves of different sizes and also to observe the impact of different amounts of force on wave height. Once they have explored wave-making, invite them to add to the growing list of "wondering" questions.
Explore ideas grounded in place	**How oyster reefs change the impact of waves on shorelines.** Introduce children to the idea of shoreline erosion by watching a brief video on living shorelines that focuses on growing back oyster reefs. Then facilitate a gallery walk of photos of different models of artificial reef structures that scientists have developed. In groups, they can replicate one of the existing designs using waterproof modeling compound (plasticine), cheesecloth or mesh bags, and tiny shells (easily found at craft stores and dollar stores). Once they build a model scaled to the width of the wave tanks and just below the water line, they can test their impact on wave height. I usually have enough groups for each design to be tested by two different groups, which helps with accuracy of data. Once they've tested existing designs, they can develop ways to improve the design in ways that they think will allow more oysters to thrive and better reduce the waves hitting the shoreline.
Define problem in need of action	**Learn about Living Shorelines and How They Help Communities Vulnerable to Storms and Sea Level Rise.** Again, starting this section with a video or photo gallery walk helps children build knowledge of how communities are being impacted by increased storms and rising water. If there is a resource that highlights efforts in or near your community, that is a great place to start. Also be aware that affluent communities are the most likely to have benefited from nature-based approaches to shoreline protection. Consider having children look at photos of living shoreline installations versus seawalls versus no protection and discussing which approach(es) are most likely to support thriving communities for humans as well as water-dwelling organisms. One way to build knowledge is to have each student groups look at a set of photos or watch a short video and then share information with the class onto a growing list of features of living shorelines.

(*Continued*)

	Finally, introduce the design challenge: imagine that a community has been damaged by storms, and they want to protect their community. They also want to protect the shoreline and the creatures living there. How might your student engineers help them design ways to protect the community and the environment and encourage families to enjoy the shoreline?
Design hope-filled actions	**Design and Test Living Shoreline Models.** I usually spend a few weeks asking families to contribute leftover materials for this engineering challenge: pebbles/ gravel, artificial plants (I'm always surprised at how easy this is to obtain when I ask!), shells, and porous fabric. I keep a classroom supply of plasticine, which is endlessly re-useable, and this serves as a base for the models because it ensures the models sink rather than float! I recommend setting a shared goal of reducing wave height by 2–3 cm, rather than treating the design phase as a contest. Student groups usually need at least two rounds of plan/ draw/ build/ test/ evaluate to develop a model that meets the goal.
Share and learn from community	An ideal culmination of this unit is a visit to a shoreline that either already has a living shoreline installation or that is being impacted by rising water and has potential as a living shoreline. If you live in a community near a shoreline, partnering with a community organization to organize a field trip where children can share their ideas and learn about local community efforts is a powerful way to connect children's classroom learning to the larger community. If you live in an inland community, consider connecting with a classroom in an area that is more directly impacted by ocean and land interactions. Classrooms connecting via videoconference can share information about how their communities live in relationship with local waterways and provide enthusiastic audiences for each others' designs and ideas.
Reflect and synthesize systems	**Dream and draw.** As a final activity, consider asking children to "dream and draw" a community that has a healthy, dynamic shoreline, based on what children have learned and designed throughout the unit. Encourage them to spend time imagining what it would be like to live in this community as a toddler, and elementary schooler, a teenager, and an adult. These dream and draw products are often beautiful repositories of children's hopes and dreams for their communities, and they may inspire continued attention to how humans and nature interact.

Recommended Children's Books

Dominguez, A. (2020). *Stella Diaz never gives up*. Roaring Brook Press.

Ferrari, S. L. (2019). *Saving tally: An adventure into the great pacific plastic patch* (G. Vallecelli, illus.).

King, H. T. (2021). *Saving American beach* (E. Holmes, illus.). G.P. Putnam's Sons Books for Young Readers.

Messner, K. (2018). *The brilliant deep: Rebuilding the world's coral reefs* (K. Forsythe, illus.). Chronicle Books.

Yezerski, T. F. (2011). *Meadowlands: A wetlands survival story*. Farrar, Straus and Giroux.

Notes

1 National Oceanic and Atmospheric Administration (n.d.). *What percentage of the American population lives near the coast?* https://oceanservice.noaa.gov/facts/population.html

2 Naddaf, R. (2020, January 20). Six ways we are all connected to the ocean. *Oceana*. https://oceana.ca/en/blog/six-ways-we-are-all-connected-ocean/

3 United States Geological Survey (2015, December 18). Streamer app. https://txpub.usgs.gov/DSS/streamer/web/

4 National Oceanic and Atmospheric Administration. (2022, August 26). *Quantifying the ocean carbon sink.* https://www.ncei.noaa.gov/news/quantifying-ocean-carbon-sink

5 Mohan, L. (2013). *One ocean: A guide for teaching the ocean in grades 3-8.* National Geographic. https://education.nationalgeographic.org/resource/one-ocean-teacher-guide

6 National Oceanic and Atmospheric Administration PMEL (n.d.) *What is ocean acidification?* https://www.pmel.noaa.gov/co2/story/What+is+Ocean+Acidification%3F

7 Monahan, P. (2018, February 23). Ocean acidification poses new concern for SF bay water. *SF State News*. https://news.sfsu.edu/news-story/ocean-acidification-poses-new-concern-sf-bay-water

8 Su, J., Cai, W. J., Brodeur, J. et al. (2020). Chesapeake Bay acidification buffered by spatially decoupled carbonate mineral cycling. *Nature Geoscience, 13,* 441–447. https://doi.org/10.1038/s41561-020-0584-3

9   Monterey Bay Aquarium (n.d.). *How seagrass and kelp support habitats' resilience in a changing ocean.* https://www.montereybayaquarium.org/stories/seagrass-kelp-help-climate-change-ocean-acidification

10 Pierre-Louie, K. (2019, December 7). World's oceans are losing oxygen rapidly, study finds. *New York Times.* https://www.nytimes.com/2019/12/07/climate/ocean-acidification-climate-change.html?action=click&module=Well&pgtype=Homepage&section=Climate%20and%20Environment

11 Lindsey, R. (2022, April 19). *Climate change: Global sea level.* National Oceanic and Atmospheric Administration. https://www.climate.gov/news-features/understanding-climate/climate-change-global-sea-level

12 Wong, K. (2021, October 14). Solving problems with eelgrass and oysters. *Bay Nature.* https://baynature.org/article/solving-problems-with-eelgrass-and-oysters/#.YWt708AHtyF.twitter

13 The Nature Conservancy (2022, April 12). *Oyster restoration in Virginia.* https://www.nature.org/en-us/about-us/where-we-work/united-states/virginia/stories-in-virginia/oyster-restoration-in-va/

14 Sherriff, L. (2021, July 5). The scientists fighting to save the ocean's most important carbon capture system. *The Washington Post.* https://www.washingtonpost.com/climate-solutions/2021/07/05/kelp-forests-destroyed-sea-urchins/?utm_campaign=wp_post_most&utm_medium=email&utm_source=newsletter&wpisrc=nl_most&carta-url=https%3A%2F%2Fs2.washingtonpost.com%2Fcar-ln-tr%2F34169ec%2F60e484959d2fda8060f9535c%2F5a408835ade4e25c8fb9afa4%2F40%2F72%2F60e484959d2fda8060f9535c

15 Bever, K. (2021, March 1). *"Run the oil industry in reverse:" Fighting climate change by farming kelp.* National Public Radio. https://www.npr.org/2021/03/01/970670565/run-the-oil-industry-in-reverse-fighting-climate-change-by-farming-kelp

16 Sisk-Hilton, S., & Ferner, S. D. (2022). Engineering the coast: An integrated set of three design challenges to explore living shorelines. *Science and Children, 59*(4). https://www.nsta.org/science-and-children/science-and-children-marchapril-2022/engineering-coast

6

Cities: Human and Natural Systems Working Together

Grown Up Science

- How are cities systems and how are they connected to natural systems?
- How do cities impact global climate change?
- What is a heat island and how does this impact human well-being?
- How is climate instability impacting cities, and how can we work to lessen these impacts?

Hope-Filled Classroom Connections

- Explore and improve classroom systems
- Improve a system to support children's well-being: safe passage to school
- Design ways to keep our playground cool
- Design storm-safe, water-smart cities
- Learn from hope-filled examples of cities adapting to and mitigating climate change in justice-based ways

Where I live in California, rain is a seasonal but unpredictable weather event. There is usually no rain from June through September, and then, from mid-fall through spring, we often have periods of deluge followed by weeks of dry weather. Children who grow up here have lots of experience with the concept of storm runoff! When an unexpected summer rainstorm came during my backyard summer science camp, the campers were eager to explore

DOI: 10.4324/9781003393535-6

the newly wet environment. Children raced leaves through the path of running water at the sidewalk's edge. Some were excited about the muddy puddles in pavement planting squares, while others stayed as far from the mud as possible. I found one child staring at a rain gutter, watching water cascade down through a gap between the gutter and downspout.

I decided to build on this interest by asking them where they think all the rainwater goes. Children had lots of ideas: into the bay, sucked up by plants, evaporated away, spit out "somewhere" by the storm drains. At this comment, one child reminded us about the stenciled "leads to bay" signs next to many of these drains, meant to discourage littering.

The next day, I showed them a very simple model of a neighborhood, constructed of foil-covered boxes to represent buildings, laid out on a large piece of foil-covered cardboard onto which I had drawn a road and a parking lot.[1] I then introduced several materials to serve as model pollutants for our pretend city block, including small pieces from a cut up plastic grocery bag, dish soap, sequins, and a sprinkling of loose soil. We added these items where children thought the pollutants they represented would be most likely to be found. Then outdoors, which was now quite dry, we propped one end of the model with a wooden block and imagined the low end headed toward the bay (which is the low point of our local geography). We used two spray bottles to "rain" all over our city and observed what happened.

After observing the mess that the simulated rainstorm made to our model, I asked the children to consider how we could re-design our model city so that water caused less damage and allowed people to use some of the rainwater. We went on a neighborhood walk to observe the city's recently installed rain gardens, rain barrels in a neighbor's yard, and the green roof on top of my children's backyard clubhouse. Then, small groups each became "experts" in different stormwater management approaches—rain barrels, rain gardens, permeable pavement, and green roofs—by watching videos and reading short articles. I created new groups with a mix of expertise, and they set to work planning their improved city models. The results were dramatic! When they eagerly sprayed water onto their newly constructed models, water flowed into their constructed rain barrels, absorbed into felt representing green roofs, and drained out of the model in the areas of permeable pavement and rain gardens (into which they had poked holes to allow water to soak into the ground). We ended this exploration by groups drawing "dream cities" with all of their ideas combined.

Cities as Systems Connected to the Earth

Cities are critical to modern human life. Eighty-three percent of the US population, and 56% of the world's people, live in or near urban areas.[2] Over thousands of years, people have created systems and structures that allow many people to live in a relatively small, condensed area. We have created ways to move water in and out of buildings; ways to turn energy into electricity and then distribute it across short and long distances; ways to grow large amounts of food in agricultural areas and then transport it to population centers; ways to educate large numbers of children. These are just a few of the complex structures that make up the system of a city.

When we discuss cities as systems, we often use the word *infrastructure*. This refers to the internal systems that hold together the city as a whole. Some of the many parts of a city's infrastructure (many of which are connected to regional, national, and global systems) include: energy systems, water supply, waste processing, public health systems, communication networks, food production and distribution, and transportation systems, and financial systems. Because components of a city's infrastructure are large, complex, and interconnected, cities are particularly vulnerable to infrastructure disruption. If one part stops functioning, it causes a cascade of other issues. For instance, if part of the electrical grid is knocked out by a storm, it impacts communication for emergency responders and residents. Water treatment and transportation is likely to be disrupted. Lack of power impacts refrigeration and distribution of food as well as the ability for many workers to complete their jobs or for stores and service providers to open. For this reason, creating resilient, climate adapted infrastructure is critical to maintaining quality of life in cities.[3]

We sometimes make a stark dividing line between "natural" and "human-made" environments, but cities and other environments constructed by humans are always deeply intertwined with the land, water, and living things that are in, under, and around these developed places. The topography of the land, whether flat plains or rocky hills, coastal or inland, impacts what types of buildings, roads, and transit systems are most practical and resilient. Access to water is a necessity, as are ways to control how it flows through the built environment. We are increasingly aware of the importance of urban green space, gardens, and trees in supporting environmental and human well-being. Cities are more sustainable when the more rural areas around them are also supported, from maintaining rich soil for farmland to engaging in sustainable forestry to provide raw materials, maintain biodiversity, and mitigate risk from fires.

Because cities are where so much of the world's population lives, they are also where many climate damaging practices are most intense, due to the huge need for energy and resources. By 2050, as much as 2/3 of the people on Earth are projected to be living in cities, so energy and water needs for cities around the world will continue to grow.[2] The need to envision and develop sustainable city systems is critical to reducing the impacts of climate change as well as more broadly supporting environments in which humans can thrive.

Features of infrastructure can impact the behavior of individuals and groups by making it easier or harder to make environmentally sustainable lifestyle choices. For instance, creating safe bike throughways increases the appeal of biking, thus increasing the likelihood that residents will own and use bikes. Larger numbers of bikers form a community that is likely to advocate for further improvements to a bike-supporting infrastructure, thus creating a "virtuous cycle" of sustainability. Anyone who has ever spent time in Amsterdam, or even seen pictures of city streets there, can see what this might look like after many years of implementation. In the elementary years, we can help children understand how different aspects of a city system are connected, and how we might improve that system to increase quality of life now and in the future.

Classroom Connections: Cities as Systems

Where Does Our Water Come From? One of the most important and meaningful ways to help children understand the world as a network of interconnected systems is to encourage them to explore where they live. While this chapter focuses on cities, exploration of the systems that support our lives and well-being can happen in any location. For instance, we might begin with a question like "How does fresh drinking water get to the water fountains in our school?" Children can draw their ideas as a starting point for exploring what their local water sources are, how the water is made safe for drinking, and how it travels to homes, schools, and businesses. Learning about waters' journey helps children understand what a valuable resource it is and how a complex system transforms water from natural sources into clean, drinkable water for our community. In some areas, it's feasible to visit a reservoir that provides water for a community, and to discuss ways to preserve and conserve the water. In my community, most of our water comes from snow melt hundreds of miles away, beyond the bounds of most field trips, but contour maps of our state

FIGURE 6.1 Exploring a contour map of California to trace where our region's water comes from.

allow students to trace the path of snow melt from mountains to the valley where it is stored and see how the rural areas that supply our water are inextricably connected to the cities where the majority of people in our state live.

Build a Better System: In Our Classroom and in Our Community. As we discussed above, features of the infrastructure can encourage or discourage sustainable action on the part of individuals and groups, such as building safe and connected sidewalks and bike lanes that enable more people to go car-free for short trips. This connection between environment and behavior is a fun concept for children to explore at the classroom level. One way to do this is to have students arrive to the classroom one day with desks all facing walls. Children will quickly notice what the arrangement makes it harder or easier to do. For instance, looking at the teacher or seeing screen projections is challenging while facing the wall, so it may be harder to understand directions. On the other hand, quiet, solitary work might be easier. Then they can try rearranging the room: but first they need to decide what behaviors they want to make easier and which ones they want to discourage. If the goal is being able to work easily in groups, then small clusters of desks facing each other might work well, while rows of separated seats facing the teacher won't be as effective. If the goal is easy movement around the classroom, then maximum spacing between seating options is important.

Once children begin to see the connection between the structure of something, in this case classroom seating, and what people are encouraged or discouraged from doing, they can think about ways to make the school infrastructure support more sustainable actions. For instance, perhaps the school wants to encourage students to separate food scraps from other garbage so that they can be composted. Should there be one large food scrap container, or should they be placed throughout eating areas? How can they be labeled or decorated to make it clear what belongs in them and what doesn't? Is there a way to make the process fun or rewarding, for instance with classroom worm bins to see the composting in action? They can apply this same sort of behavior observation and environment modification to reduce unneeded electricity use. One of my classes of third graders decided to make brightly decorated signs to post near light switches in their homes, reminding family members to turn off unneeded lights. While my caution about holding children responsible for the behavior of adults still stands, this was an action that students in this class did have some say over, unlike, perhaps, the purchase of a family car. Making and posting these signs at school and at home helped children feel like they had power to make a difference in supporting climate-friendly everyday behavior and built understanding of the link between environment and behavior.

How Cities Impact Climate Change

Cities take up only about 2% of the land on Earth, and yet because of the density of people they house, they consume over 2/3 of the energy that humans use.[4] Scientists estimate that between 50 and 80% of global greenhouse gas emissions are tied to cities and the systems that support them.[5]

In addition to the drivers of climate being concentrated in cities, so too are many of the impacts. Many climate risks are magnified in urban areas due to the concentration of people, locations (often coastal or alongside other waterways), and impact of the built environment on both heat and the movement and absorption of water. The built environment of cities can exacerbate heat waves, as we'll discuss below. Large swaths of pavement can leave storm water with nowhere to go, increasing the severity and length of flooding during extreme weather events. Cities have huge needs for power and water, and so disruptions to the systems that provide these can be disastrous, particularly for people who are not able to flee cities or pay for supplementary services.

But cities also have the potential to be centers of innovation and adaptation to address and mitigate the risks of climate change. Cities are the primary points of energy consumption, but they also generate over 80% of global income.[6] Creating climate-friendly infrastructure solutions in urban areas has the potential for massive impact on global climate health. Their population density can also be an asset, facilitating sustainable communities that rely less on greenhouse gas emitting energy sources and make wise use of limited water resources. Building codes can encourage and require sustainable design and retrofitting of older structures. Reducing the distance that resources need to be transported to reach large numbers of people can lessen fuel use and enable the use of shorter-range electrical vehicles.

The energy needs of buildings account for 28% of CO_2 emissions worldwide. Increasing the energy efficiency of old buildings and building more energy-efficient new structures is one way to reduce the carbon footprint of cities. In addition to changes in the structures themselves, electrifying the power grid and moving to renewable sources for electricity can eliminate greenhouse gas emissions from use of lights, elevators, appliances, and all the technology inside modern office buildings, retail outlets, and homes.[5]

Cities are also promising places to encourage less use of individual cars, since services are often close to homes and mass transit infrastructure already exists. Designing cities to concentrate housing in walkable areas that are near mass transit is reduces reliance on fossil fuels, especially if public transit is also transitioned to renewable energy sources. Electrifying public transit and powering it from renewable sources could prevent 250 million tons of carbon emissions by 2050.[2]

Compact growth that focuses on climate-sustaining infrastructure can both reduce the carbon footprint of cities and increase the quality of life for residents. It is easier to design mass transit and walkable service areas in compact communities. Low-carbon systems such as bike sharing and vehicle charging stations are easier to design and maintain within smaller geographical ranges. Reducing car traffic and increasing bike and pedestrian presence on city roads decreases pollution and makes streets safer for children, elders, and others who are most at risk from both pollution and street traffic. Increasing biking and walking paths has the added benefit of encouraging healthier lifestyles, and more compact development can reduce time spent commuting. Combined with greening of cities, which we'll discuss later in this chapter, climate-friendly city design often results in more human-friendly environments as well.

Classroom Connections: Safe Passages to School

Children may not have the power to directly impact building codes or the fuel used by bus fleets, but they can explore making their own community more sustainable in hyper-local ways. For instance, in many urban communities, children may live within walking distance of their schools, but traffic and other safety issues may make walking an unsafe choice. If many of the children in your school live within walking distance, collecting data on how many cars pass through nearby intersections and observing for safe and unsafe driving can introduce children to using data to inform decision making. For instance, if their data shows that one out of every ten cars does not fully stop at a stop sign, children can contact local lawmakers to suggest changes. Some communities have installed containers full of bright, hand-held flags at either end of busy intersections that are outside the range of school crossing guards. Pedestrians can grab a flag to wave as they cross the street, making them more visible to cars and encouraging slower driving. Brightly colored crosswalks and motion sensor lights embedded in crosswalks also make intersections safer. These are infrastructure improvements that children can advocate for, using their data to support their request to city leaders. Children and families may also help organize "walking buses" where families coordinate walking to school in groups led by older children or adults, thus creating safety in numbers.

Of course, many school attendance zones make walking to school unfeasible, but this approach of gathering data about a local problem that children have some power to impact can be applied to a wide variety of issues. In Chapter 8 I describe doing a school energy audit. Students can then use data on school energy use, to encourage positive changes like having a "light turner offer" job in each classroom or suggesting that thermostats be turned down a couple of degrees in winter to reduce energy use. Whether your school is in a city, part of an outlying metropolitan area, or in a rural setting, finding ways for children to first explore the systems that support their communities and then looking for ways to make small, positive changes helps build understanding as well as develop feelings of empowerment through collective action.

Cities as Heat Islands

One of the impacts of climate change is a rise in average global temperature. As of 2021, more than 350 major cities worldwide had average summer temperatures greater than 95° f. This number is expected to increase to 970 cities, with a combined population of 1.6 billion people, by 2050.[7] Extreme heat is an ongoing concern for the world's hottest areas, but it is also a problem for temperate climate cities. Many cities in which most buildings are not air conditioned due to historically mild summers are now experiencing more days of extreme heat, and they must find ways to keep residents safe, especially vulnerable groups like the elderly and those experiencing homelessness. Increasingly frequent and severe heat waves make climate-aware cooling strategies urgently important. Current air conditioning technologies increase greenhouse gas emissions, and yet people need ways to escape dangerous heat.[4]

While increased heat waves impact communities of all sizes, cities are sometimes more heavily impacted by heat waves than less densely populated or more natural environments. Cities are often referred to as "heat islands" because they are often significantly hotter during a heat wave than nearby areas with more natural habitat. Why is this the case? Built structures such as roads and buildings tend to absorb more of the sun's energy than natural features such as fields or forests. As you may have experienced if you've ever walked barefoot on pavement versus grass in summer, hardscape absorbs a lot of heat, and dark surfaces, like roads and dark roof tiles, absorb more than lighter colored surfaces. These structures also reflect this energy back into the environment, since it is not being used or transformed, as it is when it is absorbed by a plant.

The densely built nature of cities also impacts air flow. If buildings are positioned in ways that block natural wind patterns, and if they are closely spaced, they reduce the cooling effects of wind, so the heat emitted by buildings and pavements gets stuck, making the surrounding area even hotter. In addition, many of the activities that are concentrated in cities generate heat into the atmosphere, including the venting of air conditioners and heat released by vehicles.

As a result of all these factors, urban environments tend to be hotter, sometimes much hotter, than less densely populated areas around them. In the US, cities are generally 1°–7° warmer than their outlying areas during the day and 2°–5° warmer at night. Cities in areas with high humidity, where heat gets "trapped" in the water vapor that is in the air, experience even more dramatic heat island effects.[8] In addition, lower income urban neighborhoods tend to be hotter than more affluent neighborhoods due to factors including less tree cover and green space and greater density. Equitable protection of communities from the impact of heat islands is a climate justice issue.

There are a number of actions, some of them relatively inexpensive, that can reduce the heat island effect, and these solutions are both accessible to children. One strategy that can have a big impact on reducing local heat is planting street trees. Areas around street trees can have surface temperatures that are 20°–45° cooler than areas without trees![9] This is due to the shade that trees provide, both direct shade below their canopy and shade provided to buildings, reducing the solar radiation that heats them up through their windows and walls. Shading areas around homes and buildings also reduces air conditioning costs.

Trees also reduce the heat island effect through the process of evapotranspiration, by which they absorb the sun's energy and give off water into the atmosphere. This process in effect captures the sun's energy before it reaches structures such as pavement that absorb and then radiate the heat back into the atmosphere.[10] And of course, trees are also carbon sinks, as we discussed in the chapter on forests, so planting street trees has the added benefit of reducing carbon dioxide in the atmosphere.

Another way to reduce the heat island effect is to make the built surfaces of cities less heat-absorbing. Dark surfaces such as roofs and street pavement absorb more heat than light-colored surfaces. Painting roofs and pavement white allows more of the sun's energy to be reflected rather than absorbed, thus reducing ground temperatures. Widespread use of "cool roofs," including those that are light in color, reflective, or planted, has a measurable impact on daytime temperatures in cities, and pavement treatments can be even more effective, since its location causes it to release its stored heat energy at ground level into the night. In Tokyo, a test project demonstrated that installing a thermal barrier coating on roadways reduced the surface temperature by 14° f. These approaches can save money as well. In California, cool roofs save about 50 cents per square foot in energy cost versus traditional roofing.[7]

Hope-Filled Change: Cooling Our Playground

Whether or not your school is located in a city, the impact of the built environment on heat retention is readily explored by direct experimentation, and children can then apply their knowledge to solving a problem. To build understanding of the heat island phenomenon, children can investigate the difference in heat retention of different surfaces. On a sunny day, children can place dark and light tiles (I've found that hard surfaces like tiles work better than paper) in direct sunlight and measure the difference in temperature change. They can also put thermometers in the schoolyard in different areas, including areas shaded by trees, areas with grass or other natural ground cover, and paved surfaces, and then track temperature differences throughout the day. Note that in most regions of the US, results will be most dramatic when conducted at the beginning or end of the school year, times of year when temperate regions get more direct sunlight than they do during the winter, even on a sunny day.

I've found that students get excited analyzing the results of this investigation because there is an immediate action they can take by spending more time near trees and on natural ground surfaces at recess! They can also advocate for heat reduction strategies including planting more trees (or helping to care for the ones that are already there) and possibly painting paved surfaces light colors.

In addition to immediate direct action, students can imagine and share ideas about what a "perfect" cooling playground might look like. Even if they can't change the existing school playground right now, they can bring their ideas to life through drawing and modeling. For instance, they might use what they learned in measuring heat in different areas to design and build small, model shade structures, perhaps to protect a small animal figurine from the heat, and take it outside to measure how well their solutions work. They can also draw their ideas for a cooling playground, either individually or as a group mural project, where everyone contributes ideas on a large piece of bulletin board paper and then explains with labels or short written reflections. When ideas can't be implemented right away, encouraging children to dream together based on what they are learning builds their identities as people who have valuable ideas and who can work to enact positive solutions.

Cities and Water

Over half of the population of the US lives within a coastal watershed, and many others are on major waterways such as rivers. These locations have supported the growth of cities, since oceans and major rivers are trade routes that allow goods and people to move from one place to another. However, being located next to major bodies of water also makes these cities particularly vulnerable to the impacts of sea level rise and increased extreme weather events.

One of the biggest challenges for cities is addressing coastal erosion, the wearing away of coastal land due to increased wave strength, storm surges, and rising water levels. Traditionally, we have relied on *hardscape*, such as sea walls, to physically block the water from reaching coastal areas and the buildings nearby. However, these inflexible structures have some disadvantages. The most obvious is that they are limited by their size, so if the storm is bigger than the seawall is able to withstand, the water simply travels over it. In addition, seawalls create a physical barrier between habitats that evolved in relation to one another, so they often have a negative impact on coastal plants and animals.

Physical barriers are an important part of creating resilient cities, but increasingly, scientists and city planners are developing more flexible, nature-based solutions to protect coastal areas. Current coastal protection approaches often combine traditionally engineered approaches like sea walls with nature-based solutions such as living shorelines, which use restoration of wetlands and structures that support biodiversity, such as artificial oyster reefs, as more flexible forms of protection from rising sea levels. I discuss living shorelines in more detail in Chapter 5.

Of course, the oceans are not the only ways that water impacts cities. Inland flooding from rivers overflowing their bounds as well as the direct impacts of increased extremes in precipitation pose challenges to cities both because of their dense populations and because they are built of largely non-absorbant materials.[7] Again, nature-based approaches, as well as learning from traditional building design specific to a region's features, can help cities reduce the impact of storm surges. In the opening vignette of this chapter, children explored many of the strategies that can redirect and re-purpose storm water. These include replacing traditional hardscape with permeable pavement, so that water is absorbed into the ground. Green roofs, where traditional roofing materials are replaced by plants, absorb water and have the added benefit of reducing heat absorption. Rain gardens are planted areas specifically designed to absorb storm water, and like many of the solutions, they also add to quality of life by "greening" public spaces.

Strategies that allow more water to absorb into the ground not only decrease the risk of flooding, they also help restore the groundwater. As storms become more frequent as a result of climate instability, so too do periods of drought. During droughts, the underground stores of water, called groundwater, are depleted in order to maintain the farms that grow our food and to supply drinking water to communities. A sudden rainstorm does not always have as positive an effect on groundwater as we might predict, since dry land absorbs water less effectively than damp, planted areas (imagine a dry, hard sponge versus one that is already damp). Designing intentionally absorbent areas as part of a city's infrastructure helps mitigate impacts of both storms and drought.

Likewise, storing water from storms for later use is an important part of climate-resilient design. Reservoirs that collect and store water have long been a way to provide a steady supply of water to cities and towns. But smaller scale water collection can also be helpful. Installing rain barrels, which collect water from downspouts of the roofs of homes and businesses, allows stormwater to be stored for later use in watering the landscape, thus reducing municipal water use.

Finally, building design can help protect communities from the impacts of storms and rising water. Many water-adjacent communities around the world have long used storm-resilient design such as building raised homes on stilts. However, city architecture has tended to be impermeable from the ground up. Designing buildings that allow water to flow under and out can greatly reduce damage to people's homes and livelihoods.

As with inequitable impacts from the heat island effect discussed earlier, urban communities are not all equally vulnerable to storm risks. Placement of lower income communities in areas more vulnerable to storms and flooding as well as less well-maintained infrastructure mean that in many cities, those who are already most economically vulnerable also stand to lose the most. Re-designing cities to be more resilient and reduce the impact of storms provides an opportunity to approach urban design from a more justice-based framework.

Hope-Filled Change: Water-Smart Cities

Making cities water-resilient is a topic conducive to engineering design challenges at the elementary level. The opening vignette describes a fairly complex design challenge undertaken by a multi-age group (2nd-6th grade) in a setting where we had lots of time to design and build. But there are many smaller scale investigations that support children to think about, and design solutions to, the impact of water on cities. One place to start is to have children observe rain as it hits different surfaces. Where does the water from the paved part of a playground go? How do roofs prevent water from piling up on top of a building? The latter is often interesting to think about in terms of a school building, which may appear to have a flat roof, but which likely has a slightly sloped roof to allow water to flow to downspouts. They can look for places where water seems to gather rather than absorb back into the ground and then design and test ways to better deal with that water. For instance, they might test what happens on a model playground made of a single, impermeable slab (modeled with foil-covered cardboard) versus one that is made of squares of "pavement" with small gaps in between.

Also, when creation of physical models is not feasible, there is great value in an approach that I call "dream and draw" where children imagine what a solution might include and then bring this to life through drawing and describing what they envision. While drawing shouldn't take the place of all first-hand investigation, it has the advantage of allowing children to imagine ideas that they may not have the materials or skills to directly bring to life in a model. For instance, in Chapter 3, a group of students imagined a photosynthesizing car to help pull carbon out of the air. Of course, they did not have the technical knowledge to design and build such a car, but getting their idea on paper and communicating it to others helped them feel that innovative solutions are possible. Likewise, designing a dream city that is safe from storms, even if they can't build it yet, can help children develop empowered hope.

Stories of Hope-Filled Action from Cities around the World

Although national and international action to reduce the causes and impacts of climate change is not moving quickly enough, cities throughout the world are envisioning and bringing to life responses to increase resiliency. Many of these efforts also increase the quality of daily life for residents. I encourage you to learn about efforts in your own communities and to invite changemakers into your classroom to share their stories. What follow are examples of efforts in three different Latin American countries. These stories provide opportunities for children to learn about parts of the world that may be quite different from their own communities, but whose people deal with some of the same problems and address them through hope-filled action.

Reducing Heat Islands through Citywide Greening. In 2016, the city of Medellin, Colombia began the Corredores Verdes (Green Corridors) Project, an effort to plant more than 8,000 trees across the city. This has reduced the average temperature citywide by 3.6° f.[7] Increasing urban tree cover also helps preserve and bring back biodiversity in an area that has lost much of its natural, forested areas. And as we've discussed, trees serve as powerful carbon sinks, reducing the amount of greenhouse gases in the atmosphere, and they also reduce air pollution. Finally, increased exposure to plants and natural environments has positive effects on people's sense of well-being, and impacts are particularly powerful in lower income areas that previously had little access to nature.

Erosion Prevention for Interconnected Urban and Rural Communities. In San Salvador, El Salvador, a citywide effort to re-forest the hillsides that surround the city has helped to reduce landslides and also to absorb floodwater. This creates sustainable living conditions for city dwellers and also improves the livelihoods of coffee farmers whose farms are most directly impacted by the floods and landslides. This example engages the interconnected system linking cities to outlying areas and shows how addressing one issue, in this case deforestation, can improve more than one aspect of the system.[4]

Nature-Based Solutions to Create a Climate-Resilient City. As we discussed earlier, nature-based solutions to climate vulnerability are particularly promising because they reduce risks while also having other positive effects, including improved quality of life and increased support for biodiversity. The city of Xalapa, Mexico has partnered with the United Nations Environment Program to develop a multi-pronged response to increased periods of both flood and drought. They are installing rainwater collection systems on public buildings including schools in order to provide more consistent water supply. They are simultaneously working to restore and preserve the vulnerable cloud forest ecosystem that surrounds the city. This effort reduces potential damage from flooding, increases carbon capture, and provides protection for the soil and plants of nearby coffee plantations. The nature-based design project takes into account the unique, natural features of the area to design sustaining responses to the changing climate. This community is creating a more resilient and livable city while also helping reduce climate degrading actions such as deforestation and soil erosion.[6]

Hope-Filled Climate Action in Your Community

While a large percentage of children live and go to school in urban areas, of course many also live in small towns and rural areas. And as we discussed at the beginning of this chapter, cities themselves are far from uniform, and the particular climate challenges in one city may be quite different than in another. As you adapt ideas from this chapter to your context, I encourage you to learn about specific climate-based challenges in your own community and to connect with local people and groups engaging in hope-filled solutions. By connecting with people who are engaging in collective climate action in your area, you will build your own and children's connection to community while helping them see how they are part of a larger system that can lead positive change.

Justice-Based Climate Science Unit Example: Keeping Communities Safe from Storms Aligned with NGSS for Grade 3

Guiding Question: How can communities create homes that keep people safe from rising water during storms?		
Focal Disciplinary Core Idea A variety of natural hazards result from natural processes. Humans cannot eliminate natural hazards but can take steps to reduce their impacts. (ESS3.B)	**Focal Science and Engineering Practices** Constructing explanations and designing solutions Developing and using models	**Focal Cross-Cutting Concept** Cause and Effect
Engage with concept and community	**How do people build homes that protect them in the places they live?** Post pictures of houses from around the world that are designed to keep families safe during storms and rising water. Children are fascinated to see houses on stilts (Myanmar and Peru), houses that float (Netherlands), buoyant houses made from recycled materials (Vietnam), houses with raised foundations (Bangladesh and Malawi), and even a house on a raft (Peru). I suggest starting the gallery walk with silent observation and then using the prompts "I notice …" "I wonder …" and "this reminds me …" to facilitate observations and connections. If it fits with your social studies goals, consider connecting this opening activity to further exploration of homes around the world.	

(Continued)

Explore ideas grounded in place	**Introduce the design challenge**: Ask children to share what it feels like to be in a storm. If you live in a place that has been impacted by storms, children may share fears. They need space to express this and also assurance that throughout time, humans have figured out ways to build homes that fit with the land and weather around them. Ask them to imagine a home that will keep a family (or multiple families) safe in a place where there are heavy rains and floods. Give them time to draw their ideas and share with others. As they share, you can introduce terms and concepts that may be present in their drawings (for instance, water collection systems, drainage pathways, things that absorb water, filters to keep trash from entering waterways).
Share and learn from community	**Learn from community leaders**: It is important to tailor this unit to the particular issues in your community. To learn about the challenges of stormwater management and climate-resilient building in your area, consider some or all of the following: • Take a walk right after a major rainfall. Have children notice where water is going. Are there places where the water pools up that could become flooded? Can you find storm drains? Where do you think they go? Are there features of the land or of buildings that might protect structures from storms and/ or divert or store the water? • If your school is in an area with neighbors or businesses nearby, students can informally interview them during a community walk to learn how they stay safe during storms and what they think would help the community be more storm-safe. Students can also develop written surveys to share with family or other community members. • Ask students, families and community members to share memories of storms, from your community or other places they may have lived. What does it feel like during a storm? What helps people feel safe? What help do communities need to get ready for a big storm, and what help do they need afterward?
Share and learn from community	• Learn about the shelters that indigenous people in your area have traditionally built. Consider how the building materials, shape, and other features fit into the local environment and how they are (or could be) adapted to provide safety in a storm. • Ask a city planner, an architect, or a structural engineer to visit your class (in person or via videoconference) to discuss how the community is working to prepare for storms and how they are working to keep people and environments safe.

(Continued)

(Continued)

Define problem in need of action	**Build expertise:** Dividing children into expert groups allows them to build knowledge of particular techniques that divert and re-use stormwater. I have found that the following approaches are accessible to elementary students and help them understand place-based design: green roofs, permeable pavement, rain gardens, and rain barrels (see chapter text for more information). After each group has read about, watched videos, and/ or observed their system in action, they can complete an information sheet to share with others. You can then create project teams comprised of one of each "expert" so that each student can contribute different ideas to the design challenge.
Design hope-filled actions	**Design, build, and test models of storm-safe houses:** This engineering design project works well over a several day period. 1. On the first day, project teams share what they learned in their expert groups, and the team decides how they will use these ideas to make a storm-safe house. I use small plastic tubs turned upside down as the "house" to which groups add their storm-safe design elements. See sample design sheet and materials list below. 2. On the second day, introduce students to the available materials. They will work with their team to create a labeled drawing of their plan for a storm-safe house. I give each team 20 counter tokens to serve as their budget, and materials are labeled with their cost. This reduces waste and adds the realistic element of needing to make some decisions based on budget. 3. After the "chief engineer" (in the form of the teacher) approves the plan, teams will need at least one class period to bring their ideas to life. 4. Teams can then test their models by making it "rain" with a spray bottle that contains a consistent amount of water and analyzing the results (this is best done outdoors). As each team gets a turn, other classmates can admire what is effective about the model and make one or two suggestions of things they might improve in their next design. 5. Teams should then have an opportunity to improve their designs based on their results and classmates' feedback.

(Continued)

Reflect and synthesize systems	Since children can't yet bring their designs to life as full-sized buildings, one way to synthesize toward action is to engage in a classroom engineering conference. Each student group can create a poster that shows their ideas for storm-safe housing based on their design projects and their research. They can also consider issues of social justice inherent in building climate resiliency, such as: How can we make sure that everyone is safe, especially if they don't have money to build a new house? How can the water that is stored or diverted in these model buildings be shared in a way that is fair for the whole community? Are there actions you could take as children to help your community based on what you've learned (ex/ help dig a rain garden or advocate for rain barrels at your school)?

Storm Runoff House Design Challenge

Engineers:_____ **Plan#**_____

Plan: Design a house that draws water away from its foundation AND collects it for future use.

Test

	2	1	0
Foundation Protection	No water on sides or within 10 cm of house	No water on sides of house	Water runs down side of house
Water Collection	Collects more than 50% of rain for future use	Collects at least 10% of rain for future use	Collects less than 10% of rain for future use

FIGURE 6.2 Sample planning sheet for storm runoff design challenge.

Evaluate

FIGURE 6.2 (Continued).

Material	Cost
Index card	2 per token
Small cups	3 tokens each
Straws	2 tokens each
Sand	3 tokens per cup*
Gravel	3 tokens per cup*
Felt	3 tokens per piece
Craft sticks	1 token for 3
Tape	1 token for 15 cm
Foil	2 tokens for 15 cm length
Toilet paper tube	3 tokens each

FIGURE 6.3 Sample materials list for storm runoff design challenge.

Recommended Children's Books

Curtis, A. (2022). *City streets are for people* (E. Fitzgerald, illus.). Groundwood Books.

Drummond, A. (2017). *Pedal power: How one community became the bicycle capital of the world*. Farrar, Strauss, and Giroux.

Kyi, T. L. (2022). *Our Green City* (C. Larmour, illus.). Kids Can Press.

Underwood, D. (2020). *Outside in* (C. Derby, illus.). Clarion Books.

Vermond, K. (2014). *Why we live where we live* (J. McLaughlin, illus.). Owlkids.

Notes

1 Activities modified from: Engineering is Elementary Team (2013). *Don't Runoff: Engineering an urban landscape*. Boston Museum of Science.

2 United Nations (2019, September 18). *Cities: A "cause and solution" to climate change*. https://news.un.org/en/story/2019/09/1046662

3 Cutter, S. L., W. Solecki, N. Bragado, J. Carmin, M. Fragkias, M. Ruth, & T. J. Wilbanks, T. (2014): Urban systems, infrastructure, and vulnerability. In J. M. Melillo, T. C. Richmond, & G. W. Yohe (Eds.) *Climate change impacts in the United States: The third national climate assessment* (pp. 282–296). U.S. Global Change Research Program. https://nca2014.globalchange.gov/report/sectors/urban

4 United Nations Environment Programme (2022, May 6). *Around the world, cities race to adapt to a changing climate*. https://www.unep.org/news-and-stories/story/around-world-cities-race-adapt-changing-climate

5 MIT Climate Portal. (2021, March 11). *Cities and climate change*. https://climate.mit.edu/explainers/cities-and-climate-change

6 United Nations Environment Programme (2019, November 20). *Banking on nature: A mexican city adapts to climate change*. https://www.unep.org/fr/node/26774

7 Boland, B., Charchenko, E., Knupfer, S., Sahdev, S., Farhad, N., Garg, S., & Huxley, R. (2021). Focused adaptation: A strategic approach to climate adaptation in cities. *C40 Cities and McKinsey Sustainability, USA*.

8 US Environmental Protection Agency. (2022, September 2). *Learn about heat islands*. https://www.epa.gov/heatislands/learn-about-heat-islands

9 US Environmental Protection Agency. (2022, October 25). Using trees and vegetation to reduce heat islands. https://www.epa.gov/heatislands/using-trees-and-vegetation-reduce-heat-islands

10 The Conversation. (2015, December 22). *Can trees really cool down our cities?* https://theconversation.com/can-trees-really-cool-our-cities-down-44099

7

The Food We Eat and the Food We Waste

Grown Up Science	Hope-Filled Classroom Connections
• How does the global system impact human and planetary well-being? • How do food systems impact greenhouse gas emissions? • How do indigenous ways of understanding connections between land and living things connect to food justice? • What practices can mitigate the climate impacts of food production? • How does food waste impact climate change, and what practices can reduce food waste and its impacts?	• Design compost systems to reduce the impact of food waste • Explore food systems through local farmers' markets and school gardening • Learn from and advocate with indigenous peoples in your community and beyond • Learn about and share plant-based foods in our children's families and communities • Design solutions to reduce school food waste

My class was shouting and cheering. I could see the teacher next door looking at us in alarm, so I told them we would need to calm down a bit. What had gotten them so excited? Compost. More specifically, the results of our eight week

DOI: 10.4324/9781003393535-7

"compost contest." Two months ago, at the end of a unit on ecosystems and how energy flows through them, groups of four had used what they knew about decomposition to set up mason jar compost chambers. Each group received the same mass and types of food waste (part of a banana peel, bread scraps, a lettuce leaf, and part of an eggshell), and they had access to a number of other materials that they could choose to add to their jars: potting soil, wood chips, shredded paper, sand, and water. Groups designed their composting plan and shared it with me along with an explanation of what they thought would happen over the next two months and why. Then they set up the jars, including a straw through the jar lid to allow gases to escape and a thermometer to provide ongoing data about changes.

Each week, we had collected data on the jars without opening them. Groups weighed them, noted the volume by marking it on the side of the jar, measured the temperature, and drew what they could see of the contents. Now, at the end of this long-term project, we were finally opening the jars and observing the results. We did this outside for safety (in case anaerobic decomposition had produced harmful gases) and mess avoidance. After taking a final round of measurements, we emptied the jar contents one at a time into a large, clear bin. Groups rated the level of decomposition on a scale of 1-5 based on how much the food scraps had broken down or were still recognizable. While I had not set this

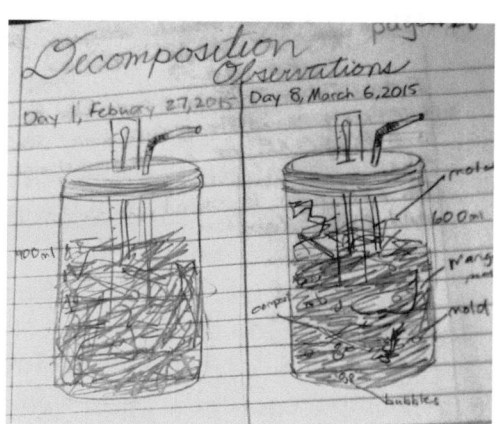

FIGURE 7.1 A student's observational drawing of their decomposition jar.

up as a contest, the students had been referring to it as such, and as each new jar was opened, they anticipated (rather loudly) how the jar's contents would compare to the previous one. Luckily, we had learned quite a bit about decomposition prior to this project, and all of the groups were successful at getting the process going in their jars.

I chose not to set this up as an experiment with a control, but if I did it again I might create some alternate jars for observation. Watching the process of food rot when surrounded by plastic waste and not exposed to air would be a dramatic comparison. It might be wise not to open such jars due to potential fumes, but it would have helped students understand why the mix of organic materials (meaning derived from living things), separated from non-biodegradable materials such as plastic waste, was critical to the composting process. As it was, students were amazed at how the food waste had so decreased in size and in some cases disappeared, and they were eager to add this (semi-)composted material to plants around the school. In our final discussion, we shared ideas about how composting could reduce greenhouse gas emissions (something they had learned about previously), how we might communicate this information to younger children and family members, and how to make sure that students were sorting their food waste into the compost bin provided by the city at our school.

Food Justice and Climate Change: A Systems Perspective

What did you eat for dinner last night? I had a pasta dish that included noodles produced in Italy, pepper from India, arugula, butter, and almonds from different parts of California, and lemons from a tree in my backyard. All except the last item were purchased at local grocery stores, so the food was transported to those stores, likely by way of warehouses that consolidated the foods prior to distribution. Before it reached the stores, the plant-based items were grown in fields, the butter started out as grass eaten by a cow, the components of the pasta were first grown on a farm and then made into the noodles at a factory before I was able to purchase them at my local store. And that's just a summary of one dinner entrée!

Our global food system allows an astonishing variety of food items to make their way onto tables around the world each day. The rise of modern agriculture has been critical to feeding a growing population. We are able to produce more than enough food to feed every person on the planet, although inequities in access and resources mean that many people worldwide and in the United States lack access to adequate food. Our current food system developed with goals of productivity and profitable distribution, and while there is great benefit for those of us who have access to this bounty, there is also a large climate cost. The global food system "from field to fork" is responsible for 21–32% of annual greenhouse gas emissions.[1] This includes every aspect of the food system, including production, transport, and disposal of food.

Learning about food and how it is connected to other global systems is directly applicable to children's lives. All children eat, and they generally care quite a lot about food. But as we explore ways to use this topic to understand climate change causes, impacts, and solutions, it's important to remember to ground the work in community action rather than individual choice. Young children generally have little power over the food choices of their families, and many of these choices are made for economic and cultural reasons, understandably not from a climate sustainability lens.

I am a vegetarian for environmental reasons. However, my personal dietary choice is rooted in privilege. I have access to a wide variety of fresh produce and enough money to purchase it. I have a job with regular enough hours that I can shop for and prepare fresh foods. I have access to resources that help me understand nutrition and develop plant-based meals that my family will eat. I also now live in a community where there are many folks who eat plant-based diets, so I benefit from a network of friends and community members with similar dietary practices. In other words, my position in economic and cultural systems allows me to make my climate-focused dietary choice. If I had a lower income or longer working hours, or if I lived in a community with fewer grocery options, I might not be able to sustain this individual choice. If more of my family traditions were based around specific foods that included meat, I might choose not to eliminate meat from my diet. And I made this change in my adult years, when I had control over the food I purchase and consume. I use my own experience as a reminder that individual choice, while important, is not a particularly helpful space in which to engage in justice-based climate action. Instead, we can focus on understanding and developing food systems that support collective action toward sustainability and equity.

Studying the interaction of human food systems and climate change brings up many issues of climate justice. As in other areas, the impact of climate change on food systems is greatest for low-income, indigenous, and other minoritized groups. Small-scale farmers, including those in indigenous communities, work with geographical limitations and within small profit margins, so a single extreme weather event has the potential to cause financial ruin. Farmworkers on larger farms are often both low income and immigrants, and their work, while critical to our nation's well-being, is economically precarious and often physically dangerous. They are exposed to health hazards caused by some industrial farming techniques, including inconsistent access to clean water and exposure to harmful chemicals in fertilizers and pesticides. Without changes to the system, these impacts will increase, and those who are closest to our nation's food production will suffer first and most.

If the food system becomes less stable, those who already experience food insecurity will be even more vulnerable, and the gap between those who can afford abundant food variety and those who struggle to have enough to live will widen. As grim as these potential outcomes are, there are also many hope-filled solutions being enacted by communities across the country that serve as models of climate-friendlier, more socially just food systems. I'll highlight these throughout the chapter, as we explore how we can better align the need to feed our whole population with the need to move away from climate-destabilizing practices.

How Food Production Impacts Greenhouse Gas Emissions

The three main greenhouse gases that contribute to anthropogenic climate change—carbon dioxide (CO_2), methane (CH_4), and nitrous oxide (N_2O)—are all emitted into the atmosphere at different points in our global systems of food production, distribution, consumption, and waste disposal. Developing ways to reduce or reverse these emissions requires people at every level of the system to work on improving distribution of healthy food while also reducing the food system's impact on global climate. Let's take a look at how each of these greenhouse gases is impacted by human food systems.

Carbon Dioxide. Some global CO_2 emissions come from agricultural machinery and the fuel needed to transport food. While moving to more locally produced foods would have a positive impact, the primary need is to move toward clean energy in the electrical grid and vehicles. We'll discuss renewable energy in the next chapter, so for now we'll look more closely at processes that are more specific to food systems.

The balance of carbon in different parts of the earth's systems is also impacted by how much and how land is used for agriculture. Clearing land for crops and pastures often involves deforestation. Reducing the number of trees, particularly large, old growth trees, means that less carbon is sequestered through photosynthesis. In addition, when soil is exposed directly to air, some of the sequestered carbon is **oxidized**, returning directly to the atmosphere. Oxidation increases when soil is tilled, exposing lower layers of soil. Farming practices that use land more efficiently can reduce the need for clear cutting, and processes such as planting cover crops and engaging in silvopasture and agroforestry, which we'll discuss below, can increase the amount of carbon sequestered in the soil and limit its release into the atmosphere.

Methane. Methane is 28 times more powerful as a greenhouse gas than carbon dioxide, and while it exists in smaller quantity than CO_2 in the atmosphere, it is the second biggest contributor to anthropogenic climate change. Methane differs from CO_2 in that it is more quickly "scrubbed" from the atmosphere, which means that lowering methane emissions would have a more immediate effect on greenhouse gases in the atmosphere. Almost one quarter of global methane emissions is the result of agricultural activity, primarily through livestock ranching. Cattle are the single biggest contributors of atmospheric greenhouse gases in the agricultural system. Pigs produce less methane than cattle, but pig farms also contribute significantly to global emissions. Food waste sent to landfills also emits methane, as we'll discuss in more detail below.

Nitrous Oxide (N_2O). This chemical accounts for approximately 7% of anthropogenic greenhouse gas emissions. While it is released in much smaller quantities than CO_2, the fact that it remains in the atmosphere unchanged for a very long time makes it a powerful contributor to global warming. According to the US Environmental Protection Agency, a pound of N_2O in the atmosphere has the same impact as 300 pounds of CO_2. The use of nitrogen-based fertilizers is the biggest human contributor of N_2O to the atmosphere. One reason for this is that fertilizer tends to be distributed in large batches, in quantities that the plants cannot fully use at the time it is applied. More strategic soil and fertilizer management can significantly decrease the amount of N_2O that makes it into the atmosphere.

How Farmers Are Helping to Mitigate and Respond to Climate Change

In recent years, many farmers have made changes that increase carbon sequestration and decrease emissions in the food production process. According to the US Department of Agriculture, soil conservation efforts have increased carbon stored in US farmland by 8.8 million tons per year.[2] This process of "carbon farming" not only reduces carbon in the atmosphere, it also improves soil health. Many of the climate-smart farming practices are accessible to children because they are visible, physical practices which can be either observed directly in farming communities or seen in photos and videos. And if you have access to, or the ability to start, a school garden, children can try some of these on a small scale. Some climate-smart practices include:

♦ *No-till farming*: Tilling, or loosening the top 6–10 inches of soil prior to planting, became typical practice in modern farming because it allowed farmers to plant more seeds, more efficiently. However, it also increases soil erosion by leaving soil uncovered by plant matter and exposed to air. Tilling can decrease soil's ability to retain water and also results in carbon being released into the air through wind erosion. In no-till farming, farmers spread out residue from the harvest and plant cover crops to prevent erosion, reduce soil impaction, and add nutrients to the soil. While switching to this method requires initial investment in equipment, long term it saves money and greatly reduces the use of fossil fuels by requiring fewer passes of farm machinery over fields.[3]

♦ *Cover crops*: These crops are grown in addition to the main cash crop, often in the off season, for the purposes of holding the soil in place and restoring soil health. Cover crops increase carbon capture (since more plants are photosynthesizing for longer periods) and help farmers adapt to destabilized and more extreme weather by reducing soil erosion and improving water flow and retention. They can also reduce the need for chemical fertilizers and lessen nutrient runoff and contamination in waterways.[4]

♦ *Precision agriculture*: This technique uses sensors in the soil to determine exactly when and how much nitrogen crops need to thrive. Applying precisely the right amount at the right time can greatly decrease nitrous oxide ending up in the atmosphere rather than in the soil and in the roots of plants.[5]

♦ *Prescribed grazing*: This practice involves ranchers moving livestock to different areas in ways that maximize food availability as well as plant health. Livestock are moved to one area while the vegetation in another is allowed to grow to a predetermined height and abundance. This improves carbon capture and soil retention due to greater volume of plants, and it can increase drought resistance as well. It also increases livestock health.[6]

◆ *Silvopasture/Agroforestry*: This refers to intentionally intermingling tree cover and pasture areas for grazing animals. As we discussed in the forest chapter, trees are critical to carbon capture in natural systems. Introducing trees to grazing land also provides shade and protection for animals and reduced soil erosion. The trees can also be used as a cash crop, for instance combining a Christmas tree farm or fruit orchard with grazing land for cows, sheep, or chicken.

◆ *Changing the diet of cattle and making use of methane*: Scientists and ranchers are experimenting with ways to reduce the amount of methane produced during cattle's digestion. Scientists at UC Davis have found that adding a small amount of seaweed to cow's diet is a promising approach.[7] Some researchers are also looking at ways to convert the methane produced by cattle into renewable fuel!

Small-Scale, Community Gardens as Part of a Climate-Friendly Agricultural System. While large-scale agriculture is the primary way that the United States and most large countries provide food for their populations, small-scale farming also plays an important role in both food security and mitigation of climate change. For instance, in areas where residents are underserved by grocery stores and markets, community gardens can provide access to healthy food and reduce food insecurity. While community gardens sometimes produce food just for the participating households, others have been developed with the purpose of selling excess produce (or eggs or in some cases even dairy products), which increases economic security for participants.

Community gardens can be built in ways that mitigate the impacts of climate change. In urban areas, garden plots can reduce the heat island effect by increasing plant cover. Areas that have been converted to gardens and small-scale farms are able to absorb water, unlike pavement, and gardens can be engineered to channel rainwater and store it for future use. Green roofs planted with food crops can also absorb a large percentage of rainwater, reducing runoff and flooding while increasing food supply. When communities grow some of their own food, they also shorten the distance food must travel from farm to table, which reduces emissions associated with transportation. Finally, the presence of community gardens encourages composting and the direct recycling of food waste back into food production.[8] Small-scale food gardens can also practice some of the same climate-friendly practices that increase carbon sequestration on large farms, including reduced tilling, cover crops, and diversification of planting rather than monoculture. School gardens, as we'll discuss below, are a wonderful way for children to directly experience the process and impact of small scale, locally tailored agriculture.

Classroom Connections: Climate Heroes at the Farmers' Market and in Your Schoolyard

There are a huge variety of factors that impact family's dietary choices, and so as we discussed in the opening to this chapter, centering individual food choices is likely to be both ineffective and potentially shaming to children. However, there is plenty that *can* be done at the elementary level to encourage children to think about where food comes from and to appreciate the role of foods lower on the food chain. One of my children came home from school in second grade telling me that he now liked salad because his class had grown and harvested lettuce, which his teacher served them topped with an "amazing dressing" that turned out to be a squeeze of lemon juice and a sprinkle of salt. He's a young adult now, but I still credit this early school gardening project with his love of salad greens. My other child is quite a gardener, a skill she started developing in her preschool, where her teacher sat out containers of soil, small spades, and watering cans and encouraged children to plant different vegetable seeds and watch them grow. Her preschool was walking distance from a weekly farmers' market held in the parking lot of a shopping center, and teachers regularly took the children there to explore the variety of food from local farmers. Both of my children, now more than a decade older, remember these experiences and credit them with helping them understand food systems.

Planting a school garden helps children connect with food systems and with natural cycles. Some schools are lucky enough to have school-wide gardens and garden instructors, and children engage in a wide variety of plant-based science and social issues over the course of their elementary years. But even a small project in a single grade level can be transformative. Many vegetables are easy and relatively quick to grow, including lettuces and other greens, carrots, and beans. Planting and tending to crops suited to your local environment provide opportunities for children to learn what plants need and how to grow food crops in ways that enrich the environment. For instance, putting mulch on top of the soil holds in moisture and reduce soil erosion, and as the mulch breaks down it adds nutrients to the soil. An added benefit is that children are often willing to eat things that they have grown that they might not otherwise be willing to try. Developing a taste for fresh vegetables alongside the knowledge needed to grow them has the potential to improve health as well as sustainable dietary practices.

Children will often want to grow foods that can't be readily grown in their location. Learning what food grows where, and how it reaches us, is a great introduction to the concept of a foodshed. If you look at the food in your own pantry and refrigerator, you'll notice that its country of origin is always labeled. It is often less clear where in the United States a particular type of food might come from, but a little internet research allows for an educated guess. One way to keep this exploration grounded in hope-filled action is to ask children to consider the foods that come from very far away, and instead of suggesting that they not eat them, instead ask them to imagine ways that they could make their way to us without fossil fuels. Could there be solar-powered ships that carry food across oceans? Could electric trains and trucks move goods across the country? Studying foodshed coordinates well with learning about transportation infrastructure and solutions that scientists and engineers are developing to make our transportation system less reliant on fossil fuels.

A field trip to a local farmers' market builds connection to your local community and lets children meet and get to know farm workers directly. People working at the farmers' market, particularly on less busy weekdays, are often eager to talk with children about where the food comes from and how it is grown. Farmers who sell their goods at these markets generally working on smaller scale, local farms that are using sustainable farming techniques. They know a tremendous amount about soil science, water conservation, and plant growth and development. If you strike up a conversation with a farmer at a market, they may be willing to talk with your class in more depth at a later time. While at the market itself, children can find foods they don't recognize, sketch them, ask what they are called, and learn about them more back in the classroom. Looking at the foods available at a farmers' market helps them understand what produce can be grown locally or regionally, versus which foods are shipped from far distances.

Another thing to explore through farmers' markets, especially for older elementary students, is who has access to this fresh, local food. Thousands of farmers' markets across the United States now accept Supplemental Nutrition Assistance Program (SNAP) benefits, making fresh, local food accessible to a wide range of community members. Of course, this is dependent on residents knowing about and being able to get to the farmers' market. Locating them near public transit hubs or in lower income neighborhoods, especially those with fewer or no grocery options, can make a huge difference in food access. Children can map out the location of local farmers' markets and then write to local government officials and farmers' market coordinators to advocate for improved access to fresh, local, climate-friendly food for all people in the foodshed.

Learning with and from Indigenous Farming Communities

Many indigenous groups have thousands of years of wisdom and experience in farming practices that make the most of available resources without depleting them. Modern farming in the United States has often rejected, or not sought out, this place-based wisdom, in favor of large-scale efficiency and uniformity of practice. One issue that is becoming more problematic as the impacts of climate change intensify is that crops have been grown for decades in places and in ways that local resources cannot support indefinitely. As climate change results in increased periods of flood and drought and other weather extremes, this becomes less and less sustainable. Matching farming practices and crops to the land and specific environmental conditions is something indigenous farming communities have always done.

Indigenous communities are being disproportionately impacted by climate instability, given the limited and often ecologically fragile areas in which tribes were forced to re-settle as their lands were stolen from them. Despite the ongoing challenges, indigenous farmers continue to use and build their knowledge to adapt to the changes as well as mitigate the causes. For example, the San Xavier Cooperative Association, developed by the Tohono O'odham Hamdag people, engages in climate-aware farm rehabilitation drawing on centuries of community knowledge. They base their work on tenets such as respect for land, elders, animals, and plants and the sacredness of water. Their practices include passive rainwater harvest, planting in accordance with characteristics of the land, and growing traditional food crops in consultation with community elders.[9]

The Ajo Center for Sustainable Agriculture, an indigenous-led food justice organization in the desert southwest, maintains a seed bank with heirloom varietals of plants specifically adapted to local environmental conditions. For example, there is a fast-maturing corn that uses much less water than modern corn crops and a bean with leaves that fold up in the sun to conserve water and survive the heat of the desert sun. Maintaining these varietals developed over thousands of years allows modern indigenous farmers to choose crops that can survive the changing conditions and also help to restore the land and sequester carbon.[10]

Indigenous knowledge and farming methods can also incorporate technological climate solutions. For example, indigenous groups who live in the Sonoran Desert have long planted crops in the shade of trees to reduce water needs and provide protections from the intensity of desert sun. Similar techniques can be used to integrate agricultural production with large solar installations, using the solar panels as a shade source for the plants.

One thing to note in this discussion is that there is concern in indigenous communities that non-indigenous groups are taking when they need something but keeping indigenous folks in economically and societally precarious positions. As children learn about the ways in which indigenous farming practices can help with climate change mitigation and adaptation, it is important to seek out stories of hope-filled action that center the voices and leadership of indigenous people. Learning about the people native to the area in which you live and learning not just about but from them, helps us get to know our community more deeply and allows children to consider issues of justice in terms of whose voices need to be heard as we work toward change.

Classroom Connections: Learning from Indigenous People in Your Community

Who were the original inhabitants of the land that your school is on? Taking the time to research the story of your community is an important first step in including indigenous perspectives and knowledge in children's experiences. In some areas, there are museums, archives, and other resources that have worked to preserve indigenous history and current experience. Even better, you can reach out to indigenous groups that still reside in your community. Many community elders and leaders are willing to take the time to speak with young people and share the stories of their people. This experience can broaden children's perspectives, helping them understand people's longstanding connection to the land and natural world as well as encouraging them to consider issues of justice around whose voices need to be amplified as we make decisions about land use and food production.

Children can also learn about indigenous farming knowledge through planting heirloom varieties of different food crops and caring for them from planting to harvest. As with all gardening projects, allowing children to actively participate in every step of the process builds scientific knowledge, increases understanding of food systems, and engages them in the hope-filled action of providing food for themselves and friends that is sustainable in the place where they live. Simultaneously learning the history of a particular food further expands their understanding of the interrelated system involving plants as food and humans as both consumers and stewards of the land we inhabit.

Climate Impact of Human Diets

Our agricultural system produces what people are willing to purchase and consume. One part of moving toward more climate-aware food practices involves changing dietary habits, especially for people in wealthy countries. Shifting dietary habits on a large scale would have a significant climate impact as well as a positive impact on people's health and well-being. In my experience, people are sometimes nervous about taking on this topic, as what we eat is both personal and cultural. Decades of body- and diet-shaming has also made food a sensitive topic. And as I described at the start of the chapter, food "choices" are tied up in issues of economic privilege. As with other aspects of bringing our planet back into climate balance, it is important to focus on actions that support community and systemic action and improved well-being for all.

Scientists measure the climate impacts of different foods in terms of their greenhouse gas emissions *intensity*. This means measuring how much greenhouse gas is emitted per calorie or per weight. When measured by weight, emissions intensity is greatest by far for beef (70 kg of greenhouse gases per kilogram of food), followed by lamb, shellfish, and cheese. At the bottom of this scale are vegetables and nuts, at.7 and.4 kg of greenhouse gas per kg of food produced. When measured by calories, the results are somewhat different. Shellfish and beef are the most greenhouse gas intensive at 26.1 and 25.9 kilograms per 1,000 kilocalories. Vegetables (a huge category!) average 3.3 kg/1000 kcal, and rice, other grains, legumes, and nuts are all under 1 kg/100 kcal.[11] Despite somewhat different "rankings" based on how it is measured, plant-based foods result in far less intense greenhouse gas emissions than animal-based foods.

People in the United States and other rich nations consume, on average, diets far higher in meat-based proteins than is sustainable from the perspective of global emissions and food security. In addition, high consumption of overly processed food has negative impacts not only on health but also on climate systems, due to the production of plastics for packaging, high use of climate-harming crops such as palm oil (which relies upon deforestation and destruction of important, natural carbon sinks in tropical areas), and impact of global distribution. Supporting sustainable and equitable dietary practices on a large scale, particularly in the world's richest nations, is a critical piece in bringing global systems into balance.

Families generally make food choices based on a combination of economic factors, food availability, cultural norms, and habits. Promoting one approach to diet, for instance veganism, as superior can lead to food-shaming, and it also ignores issues of food access, affordability, and food as an aspect of culture. "Common sense," climate-aware proposals sustainable dietary practices do not tell individuals that they must stop enjoying beef or shellfish, both of which provide significant calories and nutrients in many people's diets. Rather, we can move toward a better balance of foods. Policies that promote and provide access to a variety of well-rounded, plant-heavy diets would have a significant climate impact.

Classroom Connections: Plant-based Foods from Our Families and Cultures

One of the best ways to promote more climate-friendly dietary practices at the community level is to increase access to and enjoyment of plant-based foods. Nearly all of us can name a plant-forward food that we love and associate with home or family: beans and rice, pasta salad, dosas, fried rice, latkes, and so many others. Integrating these delicious foods into health education at the elementary level is fun as well as palate-expanding. Consider inviting parents to share a favorite plant-based dish with the class, either one by one or as part of a class potluck. Or students might write and illustrate descriptions of foods that they love that come mostly from plants.

You'll notice that school gardens come up again and again in this chapter, as they are such powerful places for learning about food systems. I mentioned earlier how my son's love of salad came directly from his experience growing lettuce at school. Sometimes merely being a part of the growing process increases enjoyment of vegetables. Young children often need many exposures to a food in order to find it palatable. In fact, researchers have found that one reason lower income families sometimes have a more restricted diet than affluent ones is that parents cannot afford the potential waste of buying a food that a child may only try a single bite of at first. Schools cannot singlehandedly solve access to affordable, fresh foods, but they can help through school gardens and school meal programs. Children who have a great experience making and eating vegetable soup with classmates may later become advocates for fairer placement of grocery stores and farmers' markets in their community.

Some school garden programs are large enough that students can bring some of the food home. In other cases, they may be able to construct small planter boxes to share with their families. Herbs are the easy to grow in small indoor spaces, and while may not play a big role in access to fresh vegetables, as with other hope-filled actions, it helps build a sense of agency in children as people who are able to produce some of their own food and have an impact on food in their community.

Food Waste and Climate Impacts

Food waste is a significant contributor to climate change that could be virtually eliminated through systems and policies that support changes in human behavior. More than a third of the food that is grown in the United States is wasted at some point in the food system. Leftover food is the largest category of waste taken to municipal landfills in the United States. All of this discarded food equates to wasting 18% of available farmland and 4 trillion tons of water in the United States alone each year. Food waste in the United States has a larger carbon footprint than the airline industry. Globally, approximately 8% of all greenhouse gas emissions can be attributed to food waste. Despite all of the food that goes to waste, tens of millions of Americans experience food insecurity.[12] The food system needs to be transformed to reduce waste at all points from production to consumption, while also more effectively distributing potentially wasted food to the people who need it.

This is an issue where individual habits, implemented at a community scale, can make a big difference. While there is waste throughout the food system, 37% of food waste happens in homes. The average American spends approximately $1300 every year on food that goes to waste. Based on a 2019 estimate, food waste costs the US economy approximately $285 billion per year, most of which is the cost to consumers of purchasing food that goes uneaten.[13]

How is food waste a climate issue? When more food is produced than is consumed, there are environmental and climate impacts at every point in the system. Using more farmland

FIGURE 7.2 The United States Environmental Protection Agency's Food Recovery Hierarchy. United States Environmental Protection Agency. *Food Recovery Hierarchy.* https://www.epa.gov/sustainable-management-food/food-recovery-hierarchy.

than needed often means that areas have been deforested, reducing natural carbon sinks. Soil is disturbed more than it needs to be, adding carbon to the atmosphere. Excess fertilizers produce more methane, and excess livestock even more. And fossil fuels are used to transport food away from farms to points of purchase, where it is then either purchased but not eaten or expires and is disposed of. With food waste, the greatest climate impacts are at the level of consumption rather than production. This is for two reasons. First, half of the waste occurs at the level of the consumer, in households and restaurants. Second, the climate impacts of food waste further along the chain from farm to consumer are cumulative, meaning that food that makes it all the way to someone's home and is then discarded has had an impact at every stage along the way. Reducing global food waste by half would reduce greenhouse gas emissions as much as taking 2570 coal-powered power plants offline. The full impact would be even greater than this, since right-sizing the amount of food produced would also reduce deforestation and soil erosion.[12]

Some promising waste prevention methods include:

♦ **Optimizing the harvest,** so that farmers have the data and economic incentives to grow only what is needed.

♦ **Improving distribution speed and efficiency,** developing technologies to better determine what is needed where, so that fresh foods don't go bad in transit.

♦ **Making better use of all parts of a food** rather than treating traditionally uneaten parts as waste. This involves developing new ways to use food byproducts in industry as well as changing consumer habits about what parts of a food are meant to be eaten.

♦ **Making changes to the human environment to encourage less waste.** For example, restaurants serving reasonable portions rather than "supersized" meals or offering smaller plates at buffets can significantly reduce consumption waste. At the household level, meal planning and making a grocery list before shopping can reduce the purchase of food that will go bad before it is eaten.

♦ **Teaching food-waste reduction as part of the curriculum.** Helping children understand the impacts of food waste and ways to reduce it allows them to play an active role in solving this issue and prepares them to adopt waste-reducing practices later in life. Teaching skills like meal planning, with connections to math, health, science, and social studies, has the potential to impact our children's health and economic well-being as well as the health of the planet.

♦ **Developing and supporting systems to redistribute food.** Many restaurants and individuals have developed ways to donate excess food to people who need it most. Sometimes there are policies originally intended to protect public health that make this challenging. Policymakers need to re-visit and revise rules to encourage food sharing while still protecting consumers, especially the most economically vulnerable.

How is food waste a climate justice issue? Despite the massive amount of waste in our food system, one out of every eight

Americans experiences food insecurity.[13] Addressing food waste includes developing systems to distribute food to those who need it rather than allow it to rot and contribute to climate change. As climate change makes crop yields less predictable, supplying enough food for everyone becomes more challenging, and when prices rise due to decreased supply, those already experiencing food insecurity suffer most. Right now, the amount of food waste produced daily contains enough calories to feed every food insecure person in the world. Better systems for minimizing food waste could have a huge impact on global climate while reducing hunger.[14]

Most food waste comes from perishable food like fresh fruits, vegetables, and meat, as well as from uneaten portions of restaurant meals. When resources are used to produce, ship, and purchase food, and then stores must dispose of uneaten food that is no longer usable, this can increase the cost of food, making it even more out of reach for economically vulnerable people. Part of reducing food waste involves better distributing the right amount of food across communities. Some of this can be accomplished from the ground up through initiatives such as community gardens and urban agriculture, but we also need larger, systemic changes that encourage wide food availability while reducing excess.

What can we do with food waste that cannot be eliminated? No matter where food waste ends up, it will decompose. *How* it decomposes has huge impacts on emissions. When food waste is mixed with non-organic trash in landfills, it breaks down through anaerobic decomposition. The easiest way to know this is happening is by the smell! In anaerobic decomposition, methane is released as a byproduct into the atmosphere. This process also does not kill harmful pathogens, so the resulting decomposed matter is potentially harmful to humans and other animals.

When food waste is instead separated from other waste and composted, the decomposition process becomes *aerobic*, meaning that it is aided by bacteria that require air. The heat generated by this process, particularly when done on a large scale, also kills pathogens, rendering the final product safe to use for gardens and farms. Composted organic material creates a material called humus, which is nutrient-rich and serves as a highly effective, natural fertilizer for crops and other plants. Aerobic decomposition does produce carbon dioxide as a byproduct, but the impact on the atmosphere is far less than that of the methane produced by food waste sent to landfills.

While preventing food waste in the first place is even more impactful than composting what cannot be consumed, improved treatment of organic/ food waste is a critical part of reducing the climate impacts of our food system. Composting can be done on a small scale in households, schools, or businesses, as startup costs do not need to be high, and maintaining a healthy compost pile is something that anyone, including children, can learn to do. There are a variety of compost solutions that work in apartments and other areas where outdoor space is limited, particularly vermi-composting, composting with the help of earthworms, which speeds up the process and reduces odor. However, larger scale composting systems are also important. Municipal composting, where organic material is collected separately from other household waste, allows communities to reduce the cost of landfills while also producing an economically important product, compost that can be used by farmers and households as well as in public natural areas. Composting is fascinating to children, and helping them understand how food decomposes, and how it can be re-purposed into valuable nutrients for plants, is a hope-filled action that they can build upon throughout their lives.

Classroom Connections: School Food Waste and Exploring Composting

Food waste is a highly accessible topic for even very young children, since they are directly involved with our food system multiple times a day! And they are easily able to see the impact of reducing food waste, making this a great project for individual classrooms or for an entire school community. Ideally, a food-waste curriculum involves learning about both reduction of food waste and composting of leftover food that cannot be avoided. Here are two ideas to get you started.

Study food waste in the school cafeteria and develop an action plan. Observe what happens during a lunch period as students dispose of their waste: is there a system in place for students to separate food waste from other trash? If so, are students doing it accurately? Students can spend one or more lunch periods observing and collecting data to inform waste-reducing actions. Older children might consider how the school community could decrease the total amount of waste AND increase composting, while younger children might just look initially at one issue or the other.

What the children find and the solutions they may be able to enact will vary based on where you live and the services available. Some communities have municipal composting, and so efforts can focus on reducing food waste and on correctly sorting what cannot be used. In areas without municipal composting, setting up school-based composting is more of an undertaking. In that case, just focusing on ways to reduce the waste is a great starting point.

Some schools have set up "share" tables where children can put uneaten items from school-provided lunches, making them available to other children who may still be hungry (this strategy involves some supervision for food safety from trained staff). Also, moving to an "offer vs. serve" approach, where children choose what is on their lunch trays, has been shown to reduce waste. Changing the lunch schedule so that children have recess before eating or are given more time to eat can also ensure that more food makes its way to growing bodies rather than into the trash. And finally, school cafeterias are allowed to donate excess food, and children can play a role in setting this up or in sorting items to be donated.[15]

Study composting and set up or improve school or classroom composting systems. The process of composting is fascinating to children and easily supported in classroom environment. Setting up and maintaining a classroom compost system is a long-term project that helps children see on a day-to-day basis how the food we eat is part of a much larger system of matter and energy cycling through different parts of the earth's system. The opening story describes an investigation by fifth graders to determine ideal composting conditions. For younger children, I recommend a worm bin, which can be set up in a large plastic tub, a discarded fish tank, or any other enclosed container. There are many resources that provide information on how to set up and maintain a worm bin.[16] While a single classroom worm bin will not be able to hold a large amount of food waste, it is a powerful learning experience that can be expanded to considering how we approach food waste on a larger scale.

As children learn about decomposition as part of the earth's system, they can advocate for composting to become part of the community's waste management system by writing to leaders or even to local media outlets. In communities that already have municipal composting, they can create posters and informational materials to remind people to sort their scraps and reduce waste.

Toward More Sustainable and Just Food Systems: Stories of Hope-Filled Action

As with all areas in which we need local and global change toward climate stability and justice, there are numerous individuals, community groups, and scientists already working toward positive change. I'll close this chapter with three examples of hope-filled action, and I encourage you to look for changemakers in your own community to partner with and learn from as you engage children in understanding food systems.

In Vermont, farmers are experimenting with a wide variety of alternate sources of food production to maintain economic security in the face of climate change. The average temperature there has risen 2 degrees, and there is 21% more precipitation than in 1900, and in winter there are increasing periods with no freezing. All of these have impacted Vermont farmers, for whom dairy cattle have been the primary agricultural product. Vermont farmers are developing a number of innovative, hope-filled actions. John Brawley has re-purposed a former dairy farm to farm shrimp, an important protein source with a much smaller climate footprint than cattle. Another family was not able to sustain their small dairy farm in the early days of the Covid-19 pandemic, and they decided to convert it to a goat farm and focus on selling goat's milk for cheese. Goat farming has a significantly lower environmental impact than cattle. Their excrement is solid, unlike cow manure, which makes it easier to compost and saves the cost and climate impact of diesel fuel needed to spread out manure. The hay that the goats eat is more easily grown locally than cow feed. Still other farmers have installed solar arrays on their land and grow saffron, a highly valued, expensive spice, around the perimeter. The panels help provide shade and protection from storms for the plants while also producing sustainable energy for the community.[17]

The indigenous-led organization Seeds of Harmony, based in Arizona, draws upon centuries of indigenous knowledge and current innovation to help communities implement water harvesting and conservation practices to continue farming on indigenous lands impacted by climate instability. They are also developing other strategies that increase food security in indigenous communities, including seed saving and support for green building projects so that community members can help one another build homes that are more comfortable and climate resilient than the mobile homes that are now common to the area but do not draw on indigenous knowledge of design in accordance with natural features of the land. The Seeds of Harmony initiative focuses on social as well as nature-based systems, working toward climate justice through community healing and sustainable practice.[18]

The Food to Share effort in the city of Fresno, California, is one of many examples of justice-based approaches to improving the food system in the face of both climate instability and economic inequity. The effort brings together donor organizations and communities with high food insecurity to create food redistribution networks, build knowledge of cooking and nutrition, and support food security and community building efforts including community gardens and "gleaning" programs to harvest unused food from household gardens. The multi-pronged effort shows how looking at the whole food system from a justice lens can result in positive climate as well as social impacts.[19]

There are food justice efforts underway in communities across the country and globally that are reducing the impacts of climate instability on food systems and increasing equitable access to healthy and abundant food for all communities. I encourage you to connect with local efforts that will allow your children to both understand local food systems and engage in hope-filled action to help their community.

Justice-Based Climate Science Unit Example: Compost and Community Aligned with NGSS for Grade 5

Guiding Question: How can we recycle the energy and nutrients in food waste to help our local ecosystems?		
Focal Disciplinary Core Idea Matter cycles between the air and soil and among plants, animals, and microbes as these organisms live and die. Organisms obtain gases and water from the environment and release waste matter back into the environment (LS2.B)	**Focal Science and Engineering Practices** Planning and carrying out investigations Developing and using models	**Focal Cross-Cutting Concepts** Energy and Matter Systems and System Models
Engage with Concept and Community	The placement of this investigation within your science curriculum will influence the starting point. I taught this toward the end of a study on energy transfer in ecosystems, so students had some understanding of photosynthesis, respiration, and movement of energy through food webs. One way to start is to set up three stations that students rotate through to engage with the idea of decomposition. One station contains a variety of food scraps that are beginning to break down (covered by an upside down, clear bin for safety). Another shows images of compost piles and bins in progress. And a third is time lapse video of composting … there are many to choose from online! At each station, students observe silently for about 2 minutes, then discuss with each other using the prompts "I notice … I wonder … This reminds me …" and then record ideas in their science notebook.	

(Continued)

(Continued)

Explore ideas grounded in place	After an initial introduction to the phenomenon of decomposition, students can learn about/ review underlying concepts in jigsaw groups. One way to do this is to provide 4–5 different readings or videos that explain: • the chemistry of decomposition, • materials that can and cannot decompose, • recommendations for balanced compost bin ingredients, • how composting can reduce the climate impact of food waste, • information on earthworms and other animal decomposers. Note: I don't use earthworms for the design challenge phase of this unit, so I usually save this reading for later. Once each "expert" group has read and summarized their reading, form jigsaw groups so that each small group has at least one person from each reading. This way, each group member brings needed and unique expertise to the design challenge.
Define problem in need of action	Introduce the design challenge: How can we design a composting system to break down food waste as quickly and completely as possible? • Explain that each group will receive the same mass of similar food scraps (I use scraps of lettuce/ leafy greens, other vegetables in small pieces, cut up banana peels, egg shell, and bread) as well as a jar to serve as their "decomposition chamber." • Other materials available for designing the chamber should include: shredded paper, wood chips/ mulch, soil, dried grass or hay, water • Groups will decide on a compost strategy and draw a labeled diagram to show the plan for their decomposition chamber
Design hope-filled actions	Next, groups build their decomposition chambers and collect data to keep track of changes over several weeks. • Using mason jars that are already marked with volume measurements makes this project easier! However, students can also do this by adding 250 ml of water at a time and marking the volume on a strip of masking tape, then emptying the jars before construction. • As student groups add materials to their jars based on their design, they can use electronic or pan balances to determine the mass of each item. They should also measure the mass of the entire set up once completed, including a piece of cardboard on top with two holes for inserting a straw (for air circulation) and a thermometer.

	• Every day for a week and then twice a week for another 6 weeks, students should measure the mass, volume, and temperature of their jars and also draw their observations. While students wait for decomposition to happen, they can learn more about composting in your community. Are there farms or community gardens nearby that produce and/ or use compost? Is there a municipal composting program for yard waste and/ or food scraps? Are there local organizations that provide worm bins or other composting systems for families who live in places without yards? As students learn about composting in your community, they can develop ideas for expanding and improving these efforts.
Share and learn from community	On the final day of the composting project, bring children outside to observe the results of each composting jar. Ask children to wear face masks to protect from potential fumes. I also recommend that only the teacher open each jar and empty it into a clear bin for observation. As each jar is "revealed" discuss the strategy the group used, what worked well, and what they might improve if they made a new decomposition chamber. If your school does not already have a composting system, creating one on a small (classroom-specific) or large (school-wide) scale is a great way to put students' knowledge to use. They will need to consider how to keep the bins healthy and low-odor, how to educate other students, and who will use the finished compost. If this is a first composting project at your school site, consider starting with a small, classroom-based system that students can use to educate others, as a school-wide system is complex to manage. If your school already composts, students can share their findings with the folks who maintain the system and suggest ways to make the system even more effective.
Reflect and synthesize systems	Help students tie this excitement of their composting knowledge to their understanding of ecosystems and food systems by doing one or more of the following synthesis activities: • Individuals or groups can make posters showing what happens to food waste when it is thrown into the trash versus when it is composted, focusing on where the energy and nutrients from this food end up. • Create a growing story starting with a seed of a food plant such as a bean or an apple tree, through the growth and development of the plant, to its use as food/ energy for animals, and then to decomposition of uneaten parts. The story can be written in small groups or as a class, and then it can be performed by each child adding on one piece of the story.

(Continued)

(*Continued*)

<table>
<tr><td></td><td>

- Students can write letters to community leaders sharing what they have learned about composting and making recommendations for increasing composting in their community.
- My fifth graders loved making things for the youngest learners at the school. They made picture books to share with kindergarteners and first graders to teach them about reducing food waste and showing them how composting works.

</td></tr>
</table>

FIGURE 7.3 Decomposition jar set up.

Recommended Children's Books

Glaser, L. (2010). *Garbage helps our garden grow: A compost story* (S. Rotner, photos). Millbrook Press.

McKenna Siddals, M. (2010). *Compost stew: An A to Z recipe for the earth* (A. Wolff, illus.). Tricycle Press.

Hillery, T. (2020) *Harlem grown: How one big idea transformed a neighborhood* (J. Hartland, illus.). Simon & Schuster/ Paula Wiseman Books.

Martin, J. B. (2016). *Farmer Will Allen and the growing table* (S. Larkin, illus.). Readers to Eaters.

Paul, B. & Paul, M. (2019). *I am farmer: Growing an environmental movement in Cameroon* (E. Zunon, illus.). Millbrook Press.

Notes

1 Lynch, J., Cain, M., Frame, D., & Pierrehumbert, R. (2021). Agriculture's contribution to climate change and role in mitigation is distinct from predominantly fossil CO_2-emitting sectors. *Frontiers in sustainable food systems*, 4: 518039. doi: 10.3389/fsufs.2020.518039

2 U.S. Department of Agriculture (n.d.). *Climate solutions*. https://www.usda.gov/climate-solutions

3 Creech, E. (2017, November 30). *Saving money, time, and soil: The economics of no-till farming*. U.S. Department of Agriculture. https://www.usda.gov/media/blog/2017/11/30/saving-money-time-and-soil-economics-no-till-farming

4 USDA Economic Research Service. (2022, June 10). *Climate change*. https://www.ers.usda.gov/topics/natural-resources-environment/climate-change/

5 Chrobak, U. (2021, June 3). *The world's forgotten greenhouse gas*. BBC. https://www.bbc.com/future/article/20210603-nitrous-oxide-the-worlds-forgotten-greenhouse-gas

6 USDA Climate Hub. (n.d.). *Rotational grazing for climate resilience in the northwest*. https://www.climatehubs.usda.gov/hubs/northwest/topic/rotational-grazing-climate-resilience

7 Quinton, Z. (2019, June 27). *Cows and climate change: Making cattle more sustainable*. UC Davis. https://www.ucdavis.edu/food/news/making-cattle-more-sustainable

8 Ackerman, K., Condard, M., Culligan, P., Plunz, R., Sutto, M. P., & Whittinghill, L. (2014). Sustainable food systems for future cities: The potential of urban agriculture. *The economic and social review*, 45(2, Summer), 189–206.

9 The San Xavier Coop. https://www.sanxaviercoop.org/about/

10 Gilber, S. (2021, December 10). Native American farming practices may help feed a warming world. *Washington Post*. https://www.washingtonpost.com/climate-solutions/interactive/2021/native-

americans-farming-practices-may-help-feed-warming-world/?i-
tid=lk_inline_manual_42

11 United Nations Climate Action. (n.d.). Food and climate change: Healthy diets for a healthier planet. https://www.un.org/en/climatechange/science/climate-issues/food

12 Kaplan, S. (2021, February 25). A third of all food in the U.S. gets wasted. Fixing that could help fight climate change. *Washington Post.* https://www.washingtonpost.com/climate-solutions/2021/02/25/climate-curious-food-waste/?utm_campaign=wp_the_optimist&utm_medium=email&utm_source=newsletter&wpis-rc=nl_optimist

13 Refed (n.d.). *In the U.S., 35% of all food goes unsold or uneaten—and most of that goes to waste.* https://refed.org/food-waste/the-problem/#overview

14 U.S. Environmental Protection Agency (2021). *From farm to kitchen: The environmental impacts of U.S. food waste.* https://www.epa.gov/system/files/documents/2021-11/from-farm-to-kitchen-the-environmental-impacts-of-u.s.-food-waste_508-tagged.pdf

15 Folliard, J., Hardy, M., & Benson, F. (2021, October 18). *Food waste in schools and strategies to reduce it.* South Dakota State University Extension. https://extension.sdstate.edu/food-waste-schools-and-strategies-reduce-it

16 Indiana Department of Environmental Management (2023). *Vermicomposting: A starter's guide for teachers.* https://www.in.gov/idem/iee/classroom-lesson-plans-and-resources/vermicomposting-a-starters-guide-for-teachers/

17 Reiley, L. & Murphy, A. (2022, December 2). *Vermont's dairy farms recede, giving way to shrim, saffron and new ideas.* Washington Post. https://www.washingtonpost.com/business/2022/12/02/vermont-dairy-climate-change-agriculture/

18 Seeds of Harmony Initiative. https://www.nativeseedsofharmony.org/

19 Bergthold, K. & Ruiz-Mendez, C. (2020). Food to share as a healthy community and environmental justice case study. California Governor's Office of Planning and Research. https://opr.ca.gov/docs/20200624-FoodToShare_Case_Studies.pdf

8

(Sustainable) Energy

Grown Up Science	Hope-Filled Classroom Connections
• What is energy and how does it change forms?	• Explore energy's transformation from one form to another
• How are fossil fuels used as energy?	• Conduct an energy audit at your school and advocate for energy-wise actions
• How do fossil fuels contribute to global climate instability?	
• What makes an energy source renewable, and how can they provide energy to power human systems?	• Transform energy from wind, water, and sun into energy for homes and vehicles
• What might a just transition to renewable energy systems look like?	• Learn from hope-filled stories of just energy transitions

It was "boat week" at my back yard science camp, an exploration of basic physics, energy transformation, and transportation in our community. In preparation for an overnight camping trip to Angel Island, a small island in the middle of the San Francisco Bay, we had spent the week learning about some of the boats that do work via the bay. In my science classrooms over the

DOI: 10.4324/9781003393535-8

years, I had often used images of the giant cargo ships that come to the Port of Oakland as a starting point for understanding buoyancy. But this time, given our upcoming trip, we zoomed in on much smaller boats: the ferries that move commuters and tourists from one side of the bay to another. We learned that over 8,000 people relied on the ferries each weekday. We also learned that while ferries, like most forms of mass transit (other than airplanes) produce lower greenhouse gas emissions per person than each passenger driving a car, most of them are still powered by fossil fuels. So as we learned the basic concepts of how boats float and how they are constructed to safely hold many people without tipping, we also experimented with alternative ways to provide energy.

Given our somewhat limited supplies, most of which I had asked families to save from their recycling bins, the results were often comical. First campers worked to create boats that moved by stored mechanical power, using rubber bands and paddle wheels. Many of the first attempts moved the wrong direction or tipped if any nearby boat was in motion, both of which were great early physics lessons! When one group tried adding pennies (the model passengers for our small boats), they saw that evenly distributing the "passengers" made the boat more stable, but it took much more energy to move the now heavier boat. After a morning of tipped boats, one camper said, "I have a feeling the ferry drivers will go on strike if they have to drive rubber band boats."

Next we learned about wind energy. There lots of sailboats in the San Francisco Bay, and a couple of the campers had even been on one. Groups were able to create sailboats that moved once they understood the importance of surface area and surface tension in creating sails that use wind energy push the whole boat. They commented on the fact that we had to use a fan to test our boats since there was no wind on the day we built these. We had a great discussion about the need for energy sources to be continuous, and this led to children bringing up batteries as things that could store energy. There were lots of questions about whether wind energy could be stored in a battery. While any form of energy could in theory be converted to stored energy in a battery, this was conceptually challenging, especially for the younger campers.

Our final boat building design challenge involved the use of small solar panels and motors, and this was by far the most exciting project of the week. We learned about a solar ferry that was being built in India (it has now been in operation for several years).[1] Looking at the images of that ferry gave campers the idea that the solar panels could also help shield passengers from rain. As children designed and tested their boats, they found connecting the small motors to the boats in ways that allowed them to move was challenging, and we needed to return to what we had learned about water wheels in the first design challenge. Eventually, though, every group was able to get their boat to move across the kiddie pool that served as our model bay. The next day, as we boarded the actual ferry that would take us to Angel Island, some of the campers eagerly pointed out the fairly flat roof and said it would be "easy" to add solar panels to make it work. As of this writing, the ferries in the San Francisco Bay have not yet converted to renewable energy, but they do provide a climate-friendlier and much more beautiful commute for many people than traveling by individual car, and our next generation of young engineers has lots of ideas for using renewable energy to make them even better!

Energy: How We Power Modern Life

What is energy? Children are most likely to have heard the term in relation to themselves, and perhaps not always in a positive way! They may think of "having a lot of energy" as a sign they aren't paying attention or aren't able to keep their bodies still during circle time. Or maybe they think of energy as what allows them to run fast at recess or why they need to eat a healthy breakfast. Whatever the connotations, most elementary children have lots of ideas about what energy is and why it is important.

The basic definition of **energy is the ability to do work**. While that definition does not make all forms of energy clear (for instance, it's not immediately clear how a lit bulb is "doing work") it can be a helpful starting point. Energy is not a "thing," meaning it is not made of molecules and does not take up space, but it does cause molecules to move and change. Many forms of energy are directly experienced in our daily lives. Some of these include:

♦ **Mechanical energy**: this refers to anything that is moving, including our own bodies! When children think of themselves in relation to energy, they are often thinking of mechanical energy. Human movement is a form of mechanical energy, but so is anything else that is moving, including vehicles, machines with moving parts, or even a dandelion seed being moved by wind through the air.

♦ **Sound energy**: Sound is caused by waves traveling through air or other matter due to the vibration of an object. For instance, when a guitar string is plucked, it vibrates, causing the air around it to also vibrate, creating sound waves that reach our ears. While we think of sound energy as traveling through air, they can also travel through solids and liquids. Placing one ear on a table and having someone tap very softly shows how easily the sound energy can move through the solid surface!

♦ **Heat/ thermal energy**: The correct term here is actually thermal energy, and heat refers to the flow of energy from a warmer to a cooler object. When molecules are heated, they speed up and collide into each other more. So faster moving molecules contain more thermal energy than slower moving ones.

♦ **Light energy** is a form of electromagnetic radiation that allows energy to be visible. The energy directly provided by the sun is in the form of light and heat, but light energy can also be created by transforming other energy forms, including burning fuel to create fire (which gives off both light and heat) and transforming electrical energy into light using a light bulb.

♦ **Electrical energy**: this type of energy is caused by the flow of electrons due to the charged nature of atoms. Humans transforming other forms of energy into usable electrical energy was one of the most important innovations of the industrial revolution.

◆ **Chemical energy**: This form of energy is stored within the bonds of molecules. For instance, living things store energy in a variety of different molecules (including glucose, as discussed in the chapter on forests) and can then transform this energy to do work in cells. Batteries store chemical energy that can be transformed into electrical energy by creating a complete circuit that allows electrons to flow through it.

Almost all energy on earth originated from the sun (geothermal and nuclear energy do not). One of the most important ideas in coming to understand the concept of energy is that it can change forms. For instance, as we discussed in the forests chapter, plants can transform energy from sunlight into chemical energy, held in the bonds of glucose molecules. When those molecules are broken down, either by the plant or when consumed by animals, that energy can be used to build and maintain an organism's structures. Wood is a fuel, able to be burned for light and heat energy, because of the stored energy in what was once a tree. And as we'll discuss below, the energy stored in plants and algae from hundreds of millions of years ago has been transformed into the fossil fuels that power so much of modern human life.

Humans have worked throughout their history to harness and use naturally occurring energy, from building fires for heat and cooking to using wind to power boats or mills to grind grain. But in the 18th and 19th centuries, there was a huge leap in our ability to transform energy and use it for work, due to the development of the engine as well as innovations to produce reliable electrical energy. This change in human's relationship with energy at the start of the industrial revolution is also the primary starting point of anthropogenic climate change. As we'll discuss below, it's important to realize that tremendous good has come out of energy innovations. Overall, humans live longer, healthier lives now than in pre-industrial times, partly through innovations in medicine and medical technologies but also because of greater ability to produce and distribute food and goods to people around the world. While the energy sources we use for this work are causing problems in our global system, we want to be careful to avoid a narrative implying that modern times are somehow "bad" and that pre-industrial times were "good."

Understanding how energy changes and transforms to power natural and human systems is a great starting point for elementary school place-based explorations. Young children may not get to the point of understanding the details of *how* energy is stored, whether in a glucose molecule or in a battery, but getting to know systems in terms of energy transfer will help build a basis for later understanding the details of energy transformation and how we might transition to more sustainable energy sources.

Classroom Connections: What Is Energy and How Does It Change?

Have you ever played with an energy stick? If you haven't, please put this book down right now and go purchase one, because I promise it's more fun than reading this![2] An energy stick is a clear plastic tube that has small light bulbs and a buzzer inside it, with wires attaching them to a battery on only one side and metal wrapped around each end of the tube. In order for a battery to work, electrical energy must be able to flow in a circuit, and when the energy stick is just sitting on a table, or being held in one hand, the circuit is "open," so nothing happens. But if you hold it with one end in each hand, suddenly it lights up and buzzes. This is because your body is acting as an electrical conductor. Energy travels from one end of the stick, through one hand, across your body to your other hand, and back into the stick. This complete circuit allows the battery to supply electrical current to the light and buzzer.

That level of detail is appropriate with upper elementary students, but even preschoolers can figure out how to make the energy stick light up and have lots of fun using it to explore the idea of energy. In a first-grade class, we first explored the energy stick as a large group. I held the stick and children made suggestions about how to get it to do something. Once they figured out a way to light it up, which took several tries, I asked them to think about other ways that might work. After several more attempts that did not light, I modeled putting one end of the stick on a metal table leg. Nothing happened (since the circuit was not yet closed). But when I also touched the table leg with my free hand, it lit up! When I did the same thing with the wooden tabletop, though, nothing happened (since wood does not conduct electricity).

I created groups of three and gave each group and energy stick of their own to explore what materials would make it light up. Partners took turns being the stick holder, the place suggester, and the safety monitor (our guidelines were not to touch or get near any electrical outlets or cords or the classroom computers and to give other groups space). After several minutes of investigation they recorded methods that had worked as well as ones that had not in their science journals. Then we returned to our circle and discussed how they thought all of this worked. Where did the energy come from that would cause the stick to light up and make noise? Some children thought it might be from inside our bodies, but one group mentioned they were able to get the stick to light up with a pipe cleaner on both ends instead of a hand. Someone said there was probably a battery inside, and another disagreed because they said that then it would just have an on/off switch. For this class, this was the opening lesson for a unit about light and sound energy, so I did not provide a full explanation, but if this was a standalone lesson, I would help them build out the battery concept. Taking apart a worn-out energy stick so that the battery inside is visible can help confirm student ideas.

What does this lesson have to do with energy related to climate change? Well, understanding what happens when fossil fuels are burned, as we'll discuss below, involves the concept of energy transforming from one kind to another. In the case of the energy stick, chemical energy is stored in the battery, and when the circuit is complete, it converts into electrical energy which can then convert into light and sound energy. Becoming comfortable with the concept of energy and how it changes forms in fun and accessible ways gets students ready to understand our much more complex systems of energy sources and usage in modern human life.

If you don't have energy sticks in your classroom, there are many other ways to explore energy changing from one form to another. Plucking a guitar string changes mechanical energy into sound energy. Our vocal cords work in much the same way, with air from our breathing providing the mechanical energy. If you encourage children to place one hand gently on their throat and feel the vibrations, they can experiment with how vibrations change with bigger and smaller amounts of energy in the form of voice volume. Hitting a tuning fork and then placing the end in water allows children to see how sound energy makes the matter around it move.

All battery-operated objects convert chemical energy to electrical energy, just as the energy stick does. Figuring out where the energy is before and after a flashlight is turned on is a great starting point for understanding energy transformation. Physical exercise converts chemical energy stored in our bodies to muscle movement, and much like understanding that batteries can store chemical energy, children can learn about how our bodies store chemical energy that we use to move, directly feeling the heat given off in the process! Once children start identifying energy all around us, they will generate endless examples as well as wonderful questions about how different machines and natural living and non-living things use or transfer energy.

What Are Fossil Fuels?

We've discussed fossil fuels throughout this book, since they are a primary contributor to anthropogenic climate change. Fossil fuels include coal, oil, and natural gas. All of these were formed from the remains of decaying plants and algae from hundreds of millions of years ago. As these remains were buried under the ground, heat and pressure transformed them. The particular environmental conditions surrounding the transformation (in swamps or under the ocean floor) determined which end product was created. However, they all store energy in essentially the same way. Because fossil fuels are the transformed remains of living things, they all contain hydrocarbons, the molecules that store energy in living things. When the fuels are burned, a chemical reaction with oxygen causes the hydrocarbon bonds to break and release energy.[3]

Coal was first identified as a fuel in China more than 4,000 years ago. Coal was also present in large quantities on the British Isles. Long before the Industrial Revolution, coal was burned, much like wood, to provide heat. Coal fire was used by blacksmiths to smelt metal, since it maintain very high heat. This same property allowed coal to power steam engines when they were developed, and this innovation led to the development of machines to power textile production (and later all manner of goods) as well as revolutionizing transportation via trains and ships powered by coal-fueled steam engines.[4]

Liquid fossil fuel gained importance with the development of the internal combustion engine used in cars and other vehicles (electric cars do not have an internal combustion engine because they don't directly burn fuel). In this type of engine, the heat from the ignited fuel creates pressure, which causes the pistons in a car's engine to move. Petroleum, a liquid refined from crude oil, is the most common fuel used in vehicles.

Natural gas has been used by humans the longest of all the fossil fuels, first as a cooking fuel since approximately 500 BCE in China. However, it has been less prevalent than coal and petroleum in much of modern industrial history. It is cleaner burning than either coal or petroleum, and so there has been some move toward replacing coal and petroleum powered technologies with natural gas. Natural gas does still emit significant carbon dioxide, though, and increased extraction of natural gas is causing other environmental problems. And like all fossil fuels, humans are using it much faster than it is produced, making it a non-renewable resource.

How Fossil Fuels Contribute to Climate Change

Fossil fuels have been a critical part of creating the modern world that we live in. They have powered our homes and vehicles as well as the factories that produce most of the goods we consume. But fossil fuel use is also the single biggest contributor to anthropogenic climate change. Approximately 75% of greenhouse gas emissions worldwide, and over 90% of carbon dioxide emissions in particular, are due to the burning of fossil fuels.[5] There are many other environmental impacts of fossil fuel use, including air pollution, impact of oil spills, and degradation of natural areas due to fuel extraction, but for this chapter we'll focus just on their direct role in contributing to global climate change.

How exactly does burning fossil fuels increase greenhouse gases in the atmosphere and ocean? You likely learned, and perhaps now teach, the difference between **potential** and **kinetic** energy. Kinetic energy is energy in action, for instance, light coming from a candle or electricity moving through an electrical circuit or your arm waving at someone. Potential energy is the stored energy that can be converted into kinetic energy to do work. As we described above, the "fuel" part of fossil fuels is due to the fact that they contain stored, chemical energy from ancient living things. In other words, fossil fuels are full of potential energy. This potential energy is stored in a type of molecule called a hydrocarbon, meaning it is made of carbon and hydrogen. There are many types of hydrocarbons made of different numbers and ratios of these two elements. In order to access the energy held in the chemical bonds of the hydrocarbons, it must be ignited in the presence of oxygen. When this happens,

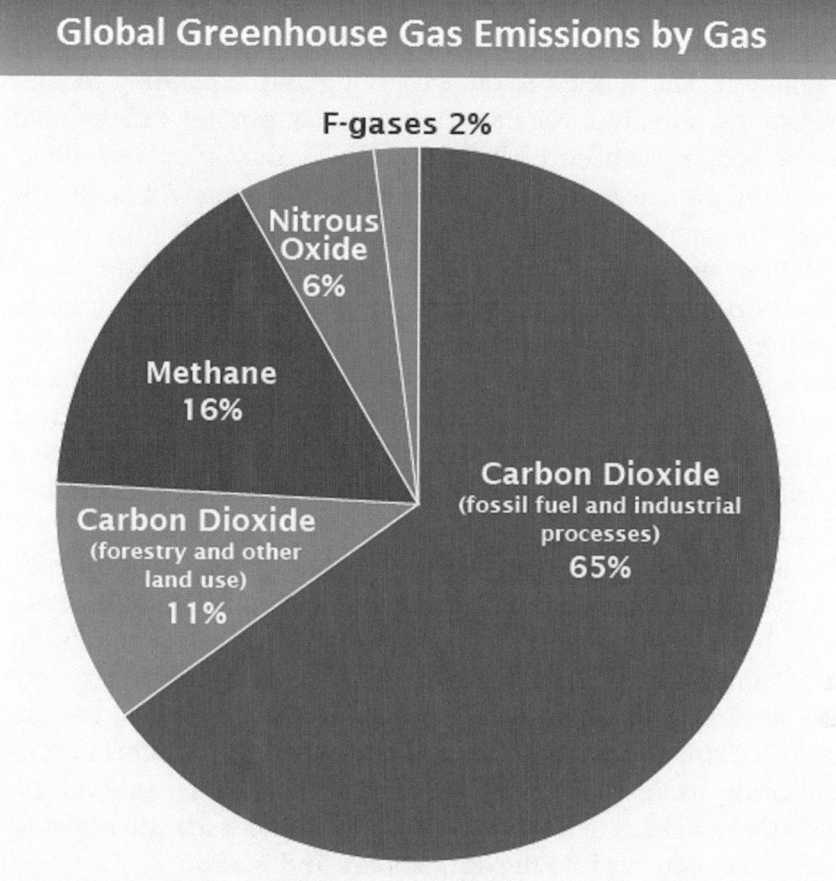

FIGURE 8.1 Sources of greenhouse gas emissions. United States Environmental Protection Agency. Global greenhouse gas emissions data. https://www.epa.gov/ghgemissions/global-greenhouse-gas-emissions-data.

the bonds are broken, releasing energy. The carbon then re-bonds with oxygen, forming carbon dioxide. So CO_2 is a byproduct of the energy releasing process.

When CO_2 from burning fossil fuels is released, about 50% of it remains in the atmosphere, where it can stay unchanged for 100 years. This increased atmospheric CO_2 is the largest contributor to human-caused global temperature rise. Approximately 25% of the CO_2 is absorbed into the ocean, and as we discussed in the oceans chapter, this is leading to ocean acidification, and the interaction between the warming atmosphere and the ocean also results in the ocean absorbing excess heat. Finally, about 25% of the CO_2 from fossil fuel use is absorbed by plants including trees and other vegetation. That's one reason preserving and expanding forests, grasslands, and other vegetated areas is so important. But it would not be possible to absorb all the excess CO_2 through plants alone. Reducing the causes and impacts of climate change will require us to radically reduce human's use of fossil fuels.[6]

Older elementary children can begin to learn the details of how burning fossil fuels results in greenhouse gas emissions, but the chemical processes at work here are generally beyond what is taught at the elementary level. Instead, we can focus on the system of energy transfer as a building block for later, fuller understanding. Children are able to easily understand that burning something releases chemicals into the air. The things they are able to see and smell in the burning process aren't actually the greenhouse gases, but understanding the process with what is directly observable prepares them to later understand how this is also going on with chemicals they can't see (and in upper grades they can use sensors to determine the presence of CO_2 and other invisible gases). For these early grades, we can help students understand the systems of energy transformations that bring usable energy to our homes, businesses, and vehicle and support them to learn about energy sources that do not emit harmful chemicals into the atmosphere and ocean.

Classroom Connections: Conducting a School Energy Audit

How much energy does your school use? Where does the energy come from? Are there ways we could use less? These are questions that children at any grade level can explore in some way, and their findings can inform school practice. Since one of the ways we can reduce greenhouse gas emissions is to use less energy, this is a potentially high impact activity, and unlike actions such as greening the energy grid, reducing wasted energy is something that children can directly help to bring about.

According to the EnergyStar program, the energy used to keep school buildings running, from lights to temperature controls to powering technology, costs more than the combined salary of all the nation's teachers! Schools that have been certified as energy efficient through the EnergyStar program use, on average, 35% less energy than typical schools. Some of this comes from changing to more efficient appliances and other upgrades that require funding, but some change is within the control of the students and teachers who use the building on a daily basis.

One way to start an elementary student-friendly energy audit is to explore the classroom and compile a list of all of the things that use electrical or gas energy. Looking for things that light up, make noise, change the temperature or are plugged in will result in a long list! This list can be the basis for a discussion about how these different technologies help us and if there are ways we could use them less to help our community and the world. Each table group could consider a different technology and present their ideas to the class, or you might work as a whole class to create a compiled set of ideas (note that not all of the ideas will be realistic or able to be implemented!)

Once students have learned about things that use energy in their classroom and have considered ways to conserve their use, they can expand their exploration to other classrooms or common areas of the school. Framing the activity as being "energy detectives" helps emphasize that students are searching for evidence and coming up with solutions to solve a problem, in this case wasted energy. The EnergyStar program provides detailed information on conducting a school-wide energy audit, but I recommend using it primarily as a teacher resource due to its complexity.[7] The Green Schools Initiative provides a more elementary-friendly worksheet that can help children notice and record information about energy use throughout the school.[8]

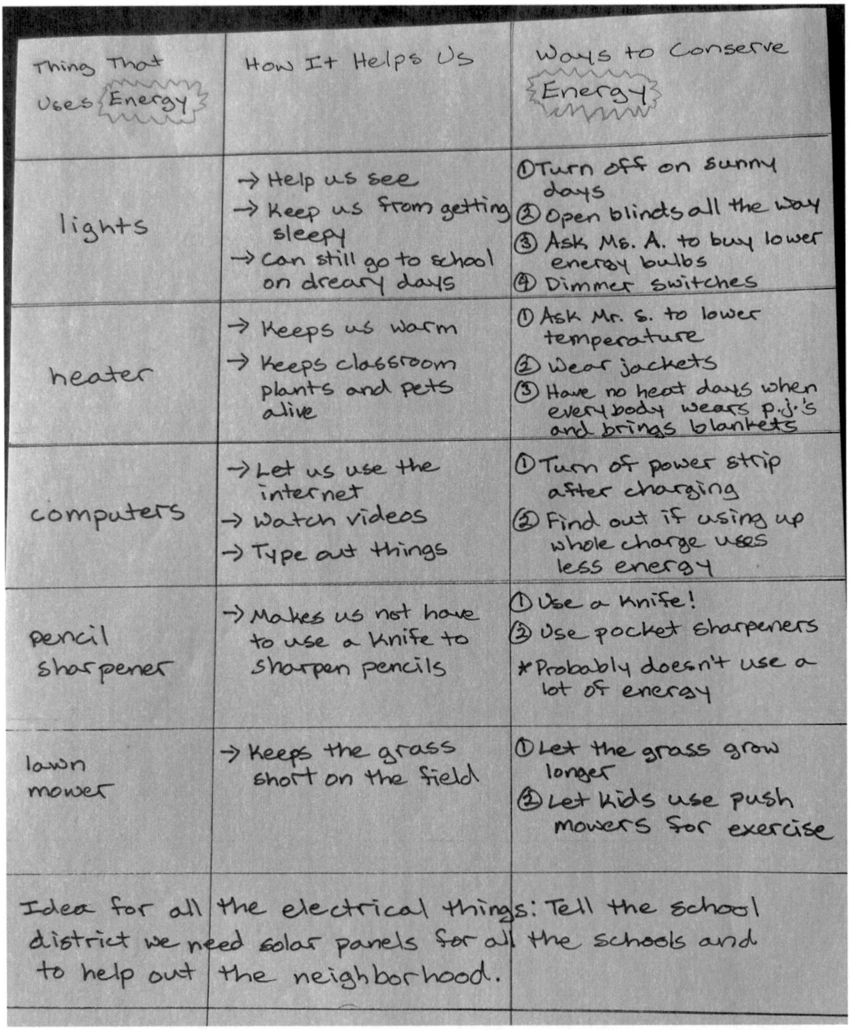

Thing That Uses Energy	How It Helps Us	Ways to Conserve Energy
lights	→ Help us see → Keep us from getting sleepy → Can still go to school on dreary days	① Turn off on sunny days ② Open blinds all the way ③ Ask Ms. A. to buy lower energy bulbs ④ Dimmer switches
heater	→ Keeps us warm → Keeps classroom plants and pets alive	① Ask Mr. S. to lower temperature ② Wear jackets ③ Have no heat days when everybody wears p.j.'s and brings blankets
computers	→ Let us use the internet → Watch videos → Type out things	① Turn of power strip after charging ② Find out if using up whole charge uses less energy
pencil sharpener	→ Makes us not have to use a knife to sharpen pencils	① Use a knife! ② Use pocket sharpeners * Probably doesn't use a lot of energy
lawn mower	→ Keeps the grass short on the field	① Let the grass grow longer ② Let kids use push mowers for exercise
Idea for all the electrical things: Tell the school district we need solar panels for all the schools and to help out the neighborhood.		

FIGURE 8.2 Start of a third grade class' energy audit and conservation brainstorming.

Inviting guests to speak about energy systems can extend children's knowledge of energy use and conservation. Your school's or district's facilities head is a great person to start with! These individuals have a systems level view of how energy is being used in your school and often have many ideas about how children can help conserve energy. Your local power company or agency may also have personnel who are eager to talk with students about energy conservation.

Once your students have gathered information and generated ideas about conserving energy, they can practice their communication skills by making posters, visiting other classrooms to explain their ideas, or making announcements over the intercom to remind other students and teachers to engage in specific, energy saving actions such as creating a "light monitor" student job or closing blinds on hot days.

What Does It Mean for an Energy Source to Be "Green?"

There is a lot of talk about moving toward more "green" energy sources as part of our national and global goals of reducing climate change. What does that mean? When people say "green energy" they are often referring to the use of energy sources that are renewable *and* do not produce any greenhouse gas emissions. However, some alternative energy possibilities *reduce* rather than eliminate emissions. In this section, we will discuss wind, solar, and water-based energy production, which do not produce greenhouse gas emissions. We'll also touch on biodiesel, which falls into the category of a "greener" potential fuel source, although it is not entirely emissions free. We won't discuss nuclear energy, as understanding its potential and the worries around it are beyond what most elementary educators will want to take on.

Solar Energy enters the earth system in the form of light and heat. While almost all energy on earth originates from the sun, when people say "solar energy," they are generally referring to energy that has not yet transformed into another form. Current technology allows solar energy to be collected and converted into electrical energy in two main ways. In one, large mirrors concentrate light energy from the sun. The resulting thermal energy heats water, which turns to steam, which can then power large electrical turbines. This type of system requires a large land area and access to reliable sunlight, so they are most often found in sparsely populated places like the Mojave Desert. You are likely more familiar with solar panels, which are increasingly common on building rooftops. Solar panels use a technology called solar voltaic cells to convert sunlight directly into electrical energy via a semiconductor.[9]

Solar energy is being used widely in the US and around the world, and in many places the price of producing electricity from solar energy has become lower than the same energy produced by fossil fuels. Solar energy is not going to run out any time soon, so it's a great example of a renewable source! Of course, solar energy technology only works when there is adequate sunlight, so there are a number of challenges to overcome, including providing steady energy at nighttime, on cloudy days, and at higher latitudes where periods of sunlight in winter are brief. Battery technology is continuing to develop, which would allow storing energy on a much larger scale. Solar energy may also be part of a system of renewable energy sources that work together to provide reliable and sufficient energy for a large population.

Wind Energy is a very accessible concept for children, as they have witnessed the power of wind to move things around the schoolyard or their neighborhood! Wind has long been used by humans to engage in work, particularly on the ocean, where predictable wind patterns allow people to harness wind via sails to move boats of a much larger size and over much longer distances than human power alone could easily accomplish. In the case of a sailboat, the mechanical energy from wind is used directly, still as mechanical energy, to move the boat. In the case of electricity production, wind's mechanical energy is captured by wind turbines, which have blades that the wind causes to spin. The blades are connected to a shaft, which is attached to a generator that converts the movement into electrical energy.

As with solar energy, a challenge of using wind for electrical energy is that in most places, wind is not always blowing at a steady rate. Figuring out ways to store and transport energy from places with reliable winds to areas with less wind is a key to making wind energy a bigger part of our energy infrastructure.[10]

Hydroelectric Power harnesses the mechanical energy of moving water and transforms it into electricity. Water-based energy has been used by humans for millennia, long before we developed the ability to use electricity. Water wheels in rivers ground grain, irrigation systems diverted the flow of water to bring it to crops, and sailors used ocean currents, often in conjunction with wind and human power, to travel across the planet. Later, during the industrial revolution, water turbines were critical to developing widespread electrical systems, such as providing power for streetlights.[8]

Large-scale hydropower today works largely due to the creation of dams that collect water at a high elevation and then release it downward, resulting in mechanical energy that can power turbines, converting the mechanical to electrical energy. Hydroelectrical power is already widely used in the US and globally. It accounts for about 7% of US and 17% of global energy production.[11]

While hydroelectric power is renewable and does not emit greenhouse gases, current hydroelectric technologies pose environmental challenges. Building dams changes the landscape, ecosystem, and water access of the region in which it is built, and dams can block or eliminate habitat for organisms such as salmon, who need to migrate from river to ocean. Also, as the climate becomes increasingly unstable and precipitation patterns less predictable, the rivers feeding dams may not be sufficient for providing power as well as irrigation. Scientists are exploring ways to harness power from ocean waves, which might be a more dependable source of energy in this time of environmental change. Of course, ocean-based technologies also have potential negative ecosystem impacts, and so scientists and communities must work together to develop technologies that allow us to use the energy from moving water without harming the ecosystems that depend upon it as well.

Biofuel is an energy source that has gained much wider use in recent years. **Biodiesel**, in particular, is created by extracting oil from plant or animal sources and converting it into a liquid fuel. Determining the environmental impact of biodiesel production and use can be somewhat complex. Like fossil fuels, biodiesel works by breaking down hydrocarbons, releasing energy stored in the bonds, and so it releases CO_2 as a byproduct. However, the entire lifecycle of biodiesel production helps "cancel out" some of these emissions due to the photosynthesis of the plants in their growing stage, which pulls CO_2 from the air. Biodiesel production also uses significant land and water resources, further complicating the environmental impacts.[12] Unless biofuel production is a significant industry in your area, it may be too complex to explore at the elementary level. However, there is one form of biodiesel production that captures children's imaginations and also has a clearer positive environmental impact: the production of fuel from used cooking oil.

In Chapter 7 we discussed the problem of food waste, which contributes significantly to greenhouse gas emissions. In addition to reducing waste, composting organic materials is an important part of making food systems more sustainable. However, cooking oil, used in large quantities around the world, does not work well in composting systems. Although cooking oil is organic material, made from either a plant or animal source, the nature of oil causes a problem in composting. Large amounts of oil will coat the other food items, creating a barrier that blocks air and moisture, both of which are necessary to the chemical process of composting. One solution is to divert the oil from the waste stream by turning it into fuel. To do this, the oil is first allowed to sit for a few days so that other food particles (and anything that isn't oil) settle to the bottom and can be more easily removed. The oil is then mixed with methanol with the aid of a catalyst, resulting in a product that can be used as liquid fuel. Biofuel from recycled cooking oil helps address the issue of waste disposal while also providing an alternative fuel source.[13]

Classroom Connections: Exploring Renewable Energy

Exploring wind, solar, and water power is fun for children and also provides an opportunity to engage in engineering and design thinking. The opening vignette provides an example of children designing and testing mechanical, wind- and solar-powered boats in the context of understanding part of our local transportation system. Overall, I have found boat design to be a little simpler than land-based vehicles only because boats don't have wheels, which can be tricky for even older elementary students to effectively attach to a vehicle of their design. If you plan to do a lot of engineering in your classroom, it is well worth it to help children understand wheels and axles, but if your time is more limited, consider using either boats or pre-made, wheeled bases for exploring energy in model vehicles.

As long as you have a fan available—or a preponderance of windy days—I've found that wind-powered vehicles are the most accessible starting point. It is helpful for children to watch videos of sailboats and other sail powered vehicles and to spend some time playing around with fabric to figure out that the sail needs to have some tension in order to "capture" the wind energy and use it to move an object. I've learned two lessons over the years to keep this design challenge child-focused but also not an exercise in frustration. First, for children in second grade and above, it is helpful to have children draw their plans first, using arrows to show how the wind energy will travel and cause the vehicle to move. For younger children, providing a sail "template" that works can be a good starting point, and then children can experiment with changes to the vehicle to make it faster, go in a straighter line, or hold more "people" (I usually use pennies as model people).

Solar vehicles require attaching a small solar panel to a motor, and the motor to something that spins in order for the vehicle to move: a rotor for boats or to the wheel structure of land-based vehicles. Again, this is a wonderful project for an engineering-focused curriculum, and one I highlight in this chapter's opening story. But if you approach it as a standalone project, children will spend a lot more time learning about motors and wheels than about the source of the energy! A simpler project that focuses more on the energy source rather than on vehicle mechanics involves creating model houses in which a very small solar panel, attached to the roof, powers an LED bulb and lights up the house. Students can experiment with putting the houses in different places and having the solar panel face different directions to figure out where the solar panels need to be placed to maximize access to the light energy from the sun. If there are buildings with solar panels in your community, you can take a walking field trip to notice where they are placed or even send small compasses home with students (another lesson, but a really fun one!) and have them determine which roofs in their neighborhood are well situated for solar panels. This also provides an opportunity to consider issues of just energy transition, discussed in the next section.

In many communities, there are active projects to collect discarded cooking oil to convert into biodiesel. One town in South Carolina offered the prize of a renovated playground to the school that helped collect the most cooking oil for re-use.[14] If there are cooking oil to fuel projects in your community, children can help publicize the efforts at school and at local restaurants and businesses in the community.

Toward a Justice-Centered Energy Transition

As we've discussed throughout this book, the impacts of anthropogenic climate change are not equitably shared by all people in every community. The impacts of fossil fuel production and use are a particularly stark example of this. The refineries that prepare fossil fuels for use by all of us are typically located in low-income communities, which suffer from greater rates of asthma, poisoned soil, and other environmental degradation than more affluent communities further from fuel preparation sites, even though less affluent people generally contribute less to climate change. Affluent nations are the highest users of fossil fuels and thus the greatest contributors to anthropogenic climate change, and yet less affluent nations are experiencing the greatest effects from increased heat and extreme weather. The causes and impacts of climate change thus far have been unjust, and responses to reduce global emissions may continue to perpetuated this inequity without a clear and determined plan for a just energy transition. According to the University of California Center for Climate Justice:

> At its core, a just transition represents the transition of fossil fuel-based economies to equitable, regenerative, renewable energy-based systems. However, a just transition is not only centered around technological change. It emphasizes employment in renewable energy and other green sectors, sustainable land use practices, and broader political economic transformations... Climate justice connects the climate crisis to the social, racial and environmental issues in which it is deeply entangled. It recognizes the disproportionate impacts of climate change on low-income and BIPOC communities around the world, the people and places least responsible for the problem.[15]

Responses to climate change that require individual economic resources are likely to perpetuate the inequitable impacts. For instance, as of this writing, electric cars are significantly more expensive than equivalent cars that run on fossil fuels, so they are still most accessible to individuals and in communities with abundant resources. As the technology matures, the price is coming down, but we must also focus on a more just transportation *system*, including improved access, reach and use of public transit so that the transportation system as a whole supports human well-being. Likewise, rooftop solar panels are an important part of greening the electrical grid, but relying on individual homeowners with resources to install them creates uneven access to clean, renewable energy. We need to consider the energy system as a whole and the impacts on marginalized communities in particular, in designing new systems that both reverse greenhouse gas emissions and provide more equitable access to reliable energy.

Children are aware of unfairness in the world. They may live in a community near an oil refinery or other high-polluting producer of energy for others. They may have experienced or know about challenges to clean water access due to environmental pollutants. They hear about oil spills that wreak havoc on marine ecosystems and pose threats to human health. Children also want to be helpers and changemakers. As we explore energy systems with them, we do not need to dwell on tragedy, but instead, we can encourage them to think in terms of community well-being and engagement rather than individual action. What efforts are going on in your community that focus on just transition? I encourage you to reach out to local organizations that are helping scientists and policy makers adopt this lens and learn about their stories and experiences. Children are eager to learn about positive change in their communities and how they can help now and as future energy innovators.

Hope-Filled Actions toward a Sustainable Energy Infrastructure

In nearly every community, there are community organizations and business/ community partnerships working to develop and implement sustainable energy infrastructure. Here are a few examples that may help identify and connect with similar efforts in your own community.

I opened this chapter with a story of my students exploring alternative energy sources for ferries. The example of the **solar ferry** already in use in Kerala, India, is exciting and shows how energy innovations can result in a cleaner and more economical transportation system. As of this writing, most ferries in the US still run on diesel fuel, but there are significant efforts underway to transition to cleaner sources. In the San Francisco Bay, most of the ferries now run on biodiesel and use advanced technology to reduce greenhouse gas emissions. In 2023, they launched a demonstration vessel that used hydrogen fuel-cell technology. As of this writing, this technology is not yet able to power large ferries, but advances in battery technology may make this possible in the future.[16] There are likely fuel conversions happening in public transit systems in or near your community, and there are often community liaisons at transit agencies who are eager to talk with elementary students about their work. They may also have developed instructional materials that you can share directly with students.

The **Wind for Schools** program, through the US Department of Energy, provides resources to partner college students with k-12 schools to design and install small-scale wind turbines (meant more for educational purposes than for large-scale energy production) on elementary and secondary school campuses. While having an actual wind turbine on your school grounds provides an outstanding, first-hand learning experience, the project also provides curricular materials to explore wind energy without an onsite turbine.[17]

The story of **William Kamkwamba**, captured in the book *The Boy Who Harnessed the Wind* shows how a young person in a small village in Malawi taught himself about electricity and wind-powered energy generation in order to build a wind turbine to provide reliable electrical power for his community. The original book was made into a child-friendly picture book. Kamkwamba's story helps children imagine how they as young people can have a powerful, positive impact on their community through identifying a problem and building their knowledge to develop innovative solutions.[18]

These stories of work toward a just energy transition show a hope-filled future, and there are many such efforts in communities across the globe. As you engage children in learning about energy, how it is generated and transformed, and how we might move toward more renewable sources, stories of the transition already in progress will help inspire both them and you to imagine what is possible when we work together toward change.

Justice-Based Climate Science Unit Example: Renewable Energy Transit Aligned with NGSS for Grade 4

Materials note: This unit requires small solar panels and motors in additional to boat building materials that can usually be collected from recycle bins. Classroom kits or individual components can be purchased through science education suppliers including Nasco.

This unit can also be easily modified to focus on land/ wheeled vehicles rather than on boats!

Guiding Questions: How do boats use energy to move?
How can we design a water-based public transportation system that uses renewable energy?

Focal Disciplinary Core Ideas	Focal Science and Engineering Practices	Focal Cross-Cutting Concept
Conservation of energy and energy transfer (PS3.B) Energy and fuels that humans use are derived from natural sources, and their use affects the environment in multiple ways. Some resources are renewable over time, and others are not. (ESS3.A)	Asking questions and defining problems Constructing explanations and designing solutions	Energy and matter

Engage with concept and community	**How do we move people across the water?** I suggest opening this study with a gallery walk of ferry boats and other people-moving boats from around the world. This will allow children who may not have traveled on a ferry to develop and contribute ideas during a class discussion. Prompts for the discussion might include: • How do the boats stay afloat when they have a lot of passengers? • How do they move? • Where do you think the energy to move them from place to place come from? • Why would a community decide to use ferry boats? The point of the initial discussion is not to definitively answer these questions, but instead to allow children to share, compare, and consider one another's ideas and for the teacher to get a sense of the knowledge and resources students bring to this study.
Explore ideas grounded in place	**How can boats use energy to move? Challenge 1: Wind Power.** I recommend an initial design challenge of building wind-powered boats that can make their way across a body of water (a small kiddie pool or large, long storage tub works well) when "wind" (from a fan) is blowing. This challenge allows children to quickly test and revise designs, learning a lot about motion and energy transfer in the process. For each round of design and testing, students can draw their groups' design, using arrows to show the movement of wind energy, and noting areas of success and needed improvements.

(Continued)

Explore ideas grounded in place	**Challenge 2: Paddle wheels.** Next, invite children to create rubber band powered boats, which demonstrates turning potential energy (in the form of the wound rubber band) into kinetic energy that moves the boat. I've found that children usually need a model of the paddle wheel structure to get them started. You can find one here: Note 1: You'll notice I've reversed the order of these challenges from what is depicted in the opening story. I've found the wind-powered boats are easier to construct and easier in terms of understanding energy transfer, so I recommend beginning with them. Note 2: Prior to these challenges, students either need some exploration of how boats float (I teach a mini-unit on how the cargo ships in our local port stay afloat, and students build clay boats), or you'll want to provide boat bases that float on their own, such as empty fruit baskets or other small food storage containers.
Define problem in need of action	At this point, I introduce some information about our local commuter ferry system: how many passengers it serves, how many people each ferry holds, and how they are powered currently (see description in chapter). I recommend providing information for boats that are relevant to your area. We then discuss: • Are the ferries good for the community? (ideas may include providing public transportation, reducing traffic and pollution in neighborhoods, allowing travelers to enjoy the natural environment of the waterway, but also potential pollution or disruption that affects the plants and animals that live in or depend on the waterway) • Are the ferries good for the environment? (ideas may include moving more people with less energy than individual vehicles, efforts to reduce emissions of engines, but also how fossil fuels release greenhouse gases) • How could people make the ferries even better for the community and the environment? Students can draw and write about their ideas for the final prompt, which they can then use in the next design challenge.
Design hope-filled actions	Now students are ready to design solar-powered ferries! I recommend spreading this out over several days to allow children time to learn new content and engage in design and revision. First, student teams can practice wiring the small solar panels in the solar boat kits to the motors. This allows for a discussion of how the panels convert solar energy into mechanical energy. They can also experiment with how much light is needed to keep the motors moving.

(*Continued*)

(Continued)

	Next, I recommend that all groups follow specific instructions to create a basic, functioning solar boat. I use the guidelines from ReCharge Labs / KidWind.[19] In my experience most elementary students learn more from making improvements to a working solar boat rather than spending too much time in the often frustrating task of "inventing" a working model. Then they can consider how to improve their boat based on criteria the class decides is important. Some of the criteria my students have come up with include: • Keeps moving as long as the sun is shining • Has comfortable places for passengers to sit (I use pennies to represent pennies) • Moves smoothly so passengers don't fall • Has some sort of covering so the motor does not harm water animals • Can change directions (most groups did not figure out how to do this, but it was a great challenge!)
Share and learn from community	The models that students develop will, of course, be much simpler than anything that could work on a large scale. But their explorations will help them understand how a solar-powered boat works. I recommend now showing some video examples of full-sized, working solar boats and asking, is there a way that solar boats could help our communities? Inviting guests from community who work with boats and/or mass transit provides an opportunity for children to connect with community efforts and share their ideas with people who are working toward sustainable transportation solutions.
Reflect and synthesize systems	As a final synthesis activity, invite students to envision what a sustainable water transportation system would look like in their community, and how they would design it to benefit the whole community. They can work together to draw and write their ideas and share them with each other and with the larger school community in the form of posters or presentations.

Recommended Children's Books

Bang, M. (2004). *My light: How sunlight becomes electricity*. The Blue Sky Press.

Bang, M. (2014). *Buried sunlight: How fossil fuels have changed the earth*. The Blue Sky Press.

Bow, J. (2016). *Energy from the sun: Solar power*. Crabtree Publishing Company.

Drummond, A. (2015). *Energy island: How one community harnessed the wind and changed their world*. Square Fish.

Drummond, A. (2020) *Solar story: How one community lives alongside the world's biggest solar plant*. Farrar, Straus and Giroux.

Heffernan, N. (2020). *Earth hour: A lights-out event for our planet* (B. Luu, illus.). Charlesbridge.

Suneby, E. (2018). *Iqbal and his ingenious idea: How a science project helps one family and the planet* (R. Green, illus.). Kids Can Press.

Notes

1 United States Environmental Protection Agency. (2023, February 15). *Global greenhouse gas emissions data*. https://www.epa.gov/ghgemissions/global-greenhouse-gas-emissions-data

2 You can find "energy stick" toys here. https://www.stevespanglerscience.com/store/energy-stick.html

3 California Academy of Sciences. (2016, August 22). *What's the deal with fossil fuels?* [Video] YouTube. https://www.calacademy.org/educators/whats-the-deal-with-fossil-fuels

4 NASA Climate Kids. (2023, March 31). *The story of fossil fuels part 1: Coal*. https://climatekids.nasa.gov/fossil-fuels-coal/

5 United Nations. (n.d.) *Causes and effects of climate change*. https://www.un.org/en/climatechange/science/causes-effects-climate-change

6 Science on a Sphere. *Ocean-atmosphere CO_2 exchange*. National Oceanic and Atmospheric Administration. https://sos.noaa.gov/catalog/datasets/ocean-atmosphere-co2-exchange/

7 United States Environmental Protection Agency. (2017) *Energy star energy efficiency student toolkit*. https://www.energystar.gov/sites/default/files/tools/K12EnergyEfficiencyStudentToolkit.pdf

8 Green Schools Initiative (n.d.). *Energy conservation checklist and tools*. http://www.greenschools.net/article.php-id=461.html

9 Office of Energy Efficiency and Renewable Energy (n.d.). History of hydropower. https://www.energy.gov/eere/water/history-hydropower

10 California Academy of Sciences (n.d.). *Optimal and sustainable: Renewable energy revamp.* https://www.calacademy.org/educators/lesson-plans/optimal-and-sustainable-renewable-energy-revamp

11 Center for Climate and Energy Solutions (n.d.). *Renewable energy.* https://www.c2es.org/content/renewable-energy/

12 MIT Climate Portal (2020, September 3). *Biofuel.* https://climate.mit.edu/explainers/biofuel

13 U.S. Department of Energy (n.d.). *Biodiesel production and distribution.* https://afdc.energy.gov/fuels/biodiesel_production.html

14 Griswold, S. (2022, October 7). *Recycle cooking oil, get a clean playground.* ABC4 News. https://abcnews4.com/news/local/recycle-cooking-oil-get-a-clean-playground

15 University of California Center for Climate Justice (2022). What is climate justice? https://centerclimatejustice.universityofcalifornia.edu/what-is-climate-justice/

16 Peter Belden, Operations Manager, Blue & Gold Fleet (personal correspondence, February 18, 2023).

17 Office of Energy Efficiency and Renewable Energy (n.d.) *Wind for schools project.* https://windexchange.energy.gov/windforschools

18 Kamkwamba, W. & Mealer, B. (2012). *The boy who harnessed the wind* (E. Zunon, illus.). Dial Books for Young Readers.

19 Instructions for solar boats and other model renewable vehicles have been developed by KidWind/ ReCharge Labs. https://www.kidwind.org/activities/recharge-labs

9

When Harm Comes to Our Communities: Teaching and Learning in the Presence of Natural and Human Disasters

Grown Up Science	Hope-Filled Classroom Connections
• How do children respond in times of environmental crisis? • What strategies support children experiencing climate anxiety?	• Support children to share fears and respond with age-appropriate information • Learn about community helpers • Engage in hope-filled helping projects to support children's agency as important members of a community

I am working with a kindergarten class, and when I tell them that I have a story to share, they all quickly gather in a circle. Several of them wonder about the bin of water and the small cups of materials sitting in front of me. I tell them they are part of the story. And then we begin.

"I'm going to tell you a story about a little girl just about your age. She lived with her Aunt and her Grandpa, and one of her favorite things to do in the whole world was to go on walks with her Grandpa. And it was his favorite thing too. When the girl got home from school in the afternoon, Grandpa's whole face would turn into a smile, he was so happy to see her. He would get her a snack, and then he would put on his jacket and get his cane from beside the door, and they would go for a walk. The girl helped her Grandpa stay steady as they walked, and Grandpa would tell the girl stories from long ago.

"Sometimes, they would walk near the little girl's school. Grandpa would remind her that he went to that very same school when he was her age. She would try to imagine Grandpa as a little boy, playing in the schoolyard. He said that when he was little, his favorite thing to do at recess was to play in the creek next to the school. He said he and his friends would take off their shoes, roll up their pants, and wade in the water looking for tadpoles and other creatures. The little girl frowned. The creek by the school was very polluted, and the teachers had warned the children not to go there. A high fence separated the playground from the creek. She asked her Grandpa, wasn't it dangerous to touch the creek water?

"So Grandpa told a story. He explained that when he was a boy, the creek was much cleaner. It was full of fish and frogs and plants, and you could find smooth, shiny rocks on the creek bed. But over time, things change. As more and more people moved to their town, they needed to build more houses and stores. They cut down the trees along the creekside. The trees helped hold the soil in place. So without the trees, when it rained, soil washed down into the creek." As I tell this part of the story, I sprinkle dirt onto the water in the bin.

"As workers built houses nearby, they would clean things up at the end of each day, but there were often piles of gravel and construction materials waiting to be used. Sometimes wind and rain would push some of the construction materials into the water." I add some gravel to the bin.

"More and more people began using the roads. When it rained, oil from the roads washed into the creek too." I add some sesame oil to the water (chosen because it shows up better than other vegetable oils).

"Sometimes, people actually left old things by the creek on purpose, since the land didn't seem to belong to anyone and they didn't have an easy way to get rid of large things they didn't need any more." I drop some paper clips and a couple of small, wooden blocks into the water and say "we can pretend like these are things like old tires or maybe even a sofa." Several children giggle at the idea of a paper clip sofa.

"Other times, children and adults might drop things and not pick them up, maybe a paper from school or a straw from a drink or a food wrapper. They didn't drop them in the creek, but wind and rain pulled some of that litter from the sidewalks into the creek too." I scatter small scraps of paper and tiny pieces of an old plastic bag onto the water.

"So the creek that used to be full of life became too dirty for fish and frogs and too dirty for people too! Some of the elders like Grandpa remembered the old creek and how beautiful and peaceful it was, but most people had forgotten it was even there. As the little girl walked with her Grandpa, she began to wonder: Could she and other kids help turn the stream back into a place where they could play?"

Finding a Safe Way into Terrifying Topics

The scenario above is not specifically about climate change. However, several of the phenomena that led to the stream in the story becoming polluted also have an impact on global climate. As we've discussed in earlier chapters, deforestation reduces plants' critical role in pulling carbon dioxide from the atmosphere into the biosphere. Traditional construction of homes and other buildings is a carbon-intensive process, and of course the cars on the road (at the time of the story) run on fossil fuels. That said, this story is more about pollution than climate impacts.

I started this chapter with this story and the design challenge it introduced because I've used it in two contexts to help children engage in knowledge-building and hope-filled action when direct climate action around environmental issues in their communities was not yet what I thought students could understand and feel powerful to enact. I first developed this story while I was working with a first-grade class in West Oakland, a community deeply impacted by environmental racism. The historically Black neighborhood around the school was once a vibrant, middle-class community, but a series of discriminatory policy decisions had negatively impacted the area over generations. These included the placement of a highway that bisected the community and exposed residents to elevated levels of air pollution, displacement of the neighborhood's business district to build an above-ground public transit station, and zoning that allowed for industrial activity near homes and schools, resulting in toxic soil and polluted air.[1] The neighborhood is adjacent to the San Francisco Bay and home to a major West Coast Port. While sea level rise is not yet visible in terms of the shoreline, it is already causing the groundwater to rise, and in the coming years this has the potential to push toxic, industrial chemicals that are currently underground to the surface, exposing this community to ever increasing levels of environmental contamination.[2]

This is a massive problem that requires municipal, state, and national resources in order for the community to address it and ensure residents' safety. The industrial contamination of this neighborhood is beyond the capacity of young children to address, but they experienced it each and every day in everything from high asthma rates to open trash piles not collected by the city to not being able to plant anything that would be eaten directly into the soil. I wanted to help them understand the connection between living and non-living things and build a sense of changemaking power in a way that felt meaningful and manageable for young children.

A couple of years later, I introduced the story and design challenge to a kindergarten class after a toxic algae bloom resulted in the closure of a lake in our regional park system that was a popular spot for family gatherings for many of the students. These toxic algae blooms have become more common in my area due to several factors attributable to climate change, most notably extended periods of drought and increased temperatures.[3]

As with the industrial pollution and impacts of sea level rise in the first school community, there was not much kindergarteners could do to directly address the algal bloom, but I wanted them to consider ways that they *could* help maintain our local waterways as healthy environments. By focusing on something that could be within the scope of influence of children—visible trash and eroded shorelines—I hoped they would better understand humans' roles in the natural systems around them and also develop their confidence as problem solvers.

Simultaneous with this kindergarten exploration, I was also working with a class of fifth graders, and because they had studied photosynthesis/transfer of matter and energy through ecosystems, they were able to more fully understand toxic algae blooms and how climate change was making them more frequent. They were able to propose and consider ideas for reducing the impacts for the lake ecosystem, including the people who enjoyed spending time there. Because their knowledge base was greater, they could take up the problem more directly and maintain and even grow their sense of agency and ability to enact meaningful change. I wanted our youngest learners to have that same opportunity, which required somewhat simplifying the problem to be solved.

Even when children are not experiencing immediate crises such as evacuation from their homes, they experience and hear about the impacts of climate change on communities. Extended drought may be impacting their ability to use water, in their homes or, in the case of farming communities, to support their livelihood. In my community we have been lucky not to have had wildfires cause extensive destruction, but almost yearly, schools are closed at least once due to poor air quality when smoke from fires elsewhere in the state blankets our region. Families in cities and towns just outside areas directly impacted by a hurricane may be called upon to help host folks who have evacuated these areas and need safe shelter. Even if children are not regularly exposed to media accounts of natural disasters caused or exacerbated by climate change, it is nearly certain that all children will be exposed in some way to these impacts.

Resources for School Communities in the Midst of a Climate Disaster

When natural disasters caused or magnified by climate change directly impact a community, schools play a critical role in maintaining children's and family's mental and physical safety. The community of educators also needs significant support as they seek to create a sense of safety for children while experiencing harm themselves. Because I am not an expert in the area of trauma response, and because there are many strong resources available that do draw on expert knowledge of supporting children and communities in the face of immediate disaster, I recommend using these and others provided by your community when there is immediate crisis.

The US Office of Human Services Emergency Preparedness and Response has compiled resources from many government agencies providing suggestions for supporting children at different developmental stages during natural disasters. I recommend this as a starting point as there is a wide range of helpful information, including specialized information for early childhood and resources developed for parents as well as educators.[4] The Center for Disease Control (CDC) has created a resource to specifically address the psychological needs of children, and it is referenced here as well.[5]

The Federal Emergency Management Agency (FEMA) has created a specialized website, ready.gov, that provides accessible resources for parents and educators to help children think about preparation for natural disasters, including preparing emergency kits and making and regularly reviewing plans for families to communicate and reunite. While even talking about the possibility of disasters with young children can seem daunting, helping them to prepare and feel "ready" often reduces fear by allowing them to feel a degree of control and knowledge of what to expect.[6]

The National Association of School Psychologists provides resources for educators to help them understand how children of different ages may respond to traumatic events as well as tip sheets for specific situations including school displacement and support for children with special needs during periods of displacement and disruption. Their resource page also provides summary posters in English and Spanish on supporting children during natural disasters that can be shared with school staff as well as families.[7]

The remainder of this chapter will provide approaches for supporting children when they are not in immediate crisis, but when the impacts of climate change reach into their world through news and through ongoing experiences.

Create Safe Ways to Process Climate Anxiety

Children (and adults) react to scary information in a variety of ways. Some will want to talk extensively about what they have experienced or heard, sometimes so much that you as a teacher may worry about its impact on other students. Some students go silent in the face of fear and need methods other than conversation to help them build meaning. Creating opportunities for children to build their knowledge as well as express their worries will not make the worry go away, but these practices can help children develop and maintain confidence in their own resiliency and the power of their community to work together toward positive change.

As I mentioned earlier, my community frequently experiences the impacts of wildfires that are often hundreds of miles away, due to wind carrying smoke and ash into the atmosphere in our region. As a result, even very young children are aware that there are big fires "somewhere," and this leads to questions about the cause and nature of the fires, concerns for people in the path of the fires, anxiety over the smoke that leads to school closures and days spent stuck indoors, and worry that their homes and families may be in danger. In situations like this, we as educators can work to balance three key factors that increase children's feeling of safety and self-efficacy:

◆ Allow children to express their feelings, questions, and wonderings
◆ Provide clear information about what is happening, particularly in direct response to children's questions
◆ Offer reassurance through gently correcting misinformation and focusing on how communities are working together to keep people and the environment as safe as possible

As tempting as it often is, telling children not to worry is a futile effort, and at worst this can cause children to hide their fears in ways that are harmful to themselves and may come out as behavioral or academic issues. Instead, we can help children see school as a place where they are allowed to share questions and fears, and where the teacher will help them self-regulate through a focus on understanding both the situation and themselves. In the next section I share one method I have used to provide space for students to learn about and process difficult events related to climate change.

Classroom Connections: Draw, Discuss, Respond

Learning about causes, impacts, and responses to natural disasters that may impact your community or others nearby can be an empowering process for children if approached sensitively. A sense of knowledge and preparedness tends to calm fears. So too does allowing children to voice their concerns and then supporting them to better understand the things they fear and to learn about ways communities are working to reduce harm from natural disasters. Collecting children's stories and questions is a critical starting point.

Drawing is often a supportive way in to discussing topics that may feel scary to talk about. For instance, if children have friends and family members who have been impacted by a hurricane, you might want to give them some quiet time to draw what they know about hurricanes or the experiences of the impacted people they may know. Sharing these drawings (by choice) can be the start of a community circle conversation. During the conversation, the teacher can largely serve as facilitator, encouraging children to share and recording questions that might guide future lessons. However, this is also a time when teachers can clarify misconceptions and focus on hopeful responses, both of which help to calm fears. For instance, if children are afraid that they might wake up in the middle of the night to a hurricane near their home, you can reassure them that scientists can predict hurricanes in advance, and that families almost always have time to evacuate, so that even if their homes are damaged, they are able to stay safe. You can highlight stories of people helping each other during the hurricane, such as neighbors checking to be sure that elders have safely evacuated and that no pets have been left behind. The initial conversation should center children's experiences and wondering, but providing calm information and comforting stories is important to maintaining children's sense of safety.

With older elementary students, you might consider using a writing/ drawing on the walls activity (sometimes referred to as "chalk talk") as the first step in this process. This involves posting a series of prompts on pieces of butcher paper throughout the room. For five to ten minutes, students can circulate freely around the room, writing or drawing on the papers in response to the prompts. I ask that students complete this process in silence so that all communication happens through written and drawn modalities. I encourage children to but stars or other symbols next to classmate's responses that resonate with them, and also to draw lines where they see connections. While I've used this instructional technique throughout my career, I first used it as a way to gather and process responses to hard topics when one of my fifth-grade classes was starting a unit on health problems in our community, many of which connected to issues of environmental injustice that their community experienced. The prompts were simple: memories of being sick, how illness has impacted my family, and questions I have about illness. The personal stories, questions, and worries shared during this time helped guide our study of community health, health equity, and the connection between health and environmental justice.

There may be people in children's families who can provide expertise from experience with past natural disasters. For example, if your class is discussing hurricanes and you live near an area impacted by the increasing strength and frequency of these storms, there are likely parents and grandparents who can talk about how their community has worked together to respond to past storms and what they are doing to prepare for the future. Inviting children to interview elders and share their experiences is another way to reduce fear through building community knowledge.

Provide Information That Focuses on Stories of Hope

Images are powerful tools for building understanding. I often use a gallery walks as the starting point for a unit so that children can look at images of the topic we will be exploring as a way to anchor our early explorations. But the power of images also means we need to be cautious and selective in how we present information about natural disasters to children regardless of age. For instance, showing a photo of an apartment building that is about to tumble into the ocean because of increased flooding and erosion of coastal areas is terrifying. However, a photo showing engineers and community members working to restore wetlands and other "softscape" that lessens the impact of coastal erosion is hopeful and action-oriented. I use this example because a few years ago, a fourth grader brought this exact photo (of the building almost collapsed into the ocean) to class because he had seen the photo in the news and connected it to our study of erosion. While the child who brought the photo to class was more excited at the curricular connection than scared of the implicit outcome, many of his classmates began asking questions that revealed worry for the folks that lived in that building as well as the safety of their own homes, even though our city was several miles away from the ocean.

In this case, I was not prepared to offer detailed information on the spot, so I spent some time helping students find their neighborhoods on a map so they could see our distance from the ocean, and then I prepared materials for the following week (I only met with this group once a week). I learned that the community in question was developing a plan for managed retreat, helping residents in now dangerous areas to move further inland, while also working to restore wetland areas to create a buffer between the ocean and land habitats.[8] I brought in lots of pictures of the restoration efforts and the people engaged in them. These images did not "undo" the frightening vision of a building about to fall into the sea, but seeing the plan to make things better in action and knowing that folks had been safely relocated from the precarious building helped children understand what was happening without focusing on the disastrous part that was well beyond their control.

Acknowledging natural disasters when they happen is important, and so too is monitoring how children are exposed to media coverage that may add to their fears and sense of power-lessness. While we cannot control what children see and hear outside of school, we can help to counterbalance disaster-filled imagery with a focus on hopeful action, and stories of communities working together to help one another in times of crisis.

Amplify the Helpers

In times of crisis, a network of people in official and unofficial capacities come together to offer immediate aid and long-term repair in communities impacted by a natural disaster. Focusing on these helpers is one way to foster feelings of hope, and it is also an opportunity to amplify jobs and roles that make a positive difference in human's responses to climate change. Think of all the stories of firefighters who put themselves in harm's way to make sure elders, children, and even pets are safely evacuated during wildfires. And stories of people who travel around by boat in submerged neighborhoods after a hurricane, pulling neighbors from rooftops and getting them to safety. And communities that band together to clean up shorelines when an oil spill brings devastation to ocean and shore ecosystems.

As you work to help children see hope in the work of so many individuals and groups in the community, be aware of whose stories are most often amplified and whose are less likely to be heard, valued, or acted upon. Consider ways you can share stories from communities that experience marginalization and who often bear the greatest burdens in climate disasters. For instance, in areas where increased periods of drought are leading to more frequent and intense wildfires, indigenous communities have wisdom from thousands of years of living in relation to the land. Indigenous people including the Yurok, Karuk, and Hoopa tribes in Northern California are leading efforts to return to cultural burning practices to better protect land from massive fires and to reclaim their right to inhabit and make decisions about use of their native land.[9] In rural farming communities, farmworkers are collaborating with scientists, using participatory action research to investigate the impact of climate change, particularly increased heat and drought, on community well-being, and they are publishing their findings and advocating for just changes to policy and farming practices.[10] What are the stories of climate response in your community that can help children engage with issues of climate justice in hope-filled ways?

As children learn these stories, they build positive images of communities as places where people work together and help each other. Of course, sadly, this is not always true, but I believe that helping children see the positive power of community and the good that comes from offering help to others makes it more likely that they will become adults who model their actions on this approach. We need caring educators, firefighters, park rangers, utility workers, policy makers, scientists, and all manner of other community members as we continue to take up the causes, impacts, and responses to climate change. Stories of helpers provide inspiring role models for our youngest generation as they grow into the leaders our planet needs.

Facilitate Hope- and Empathy-Filled Action

Finally, as we have re-visited throughout this book, engaging in hope-filled action is a powerful antidote to feelings of anxiety and powerlessness. When children are processing new of natural disasters, models of hope-filled action like the one with which I opened this chapter can be helpful. But so too are direct actions to provide assistance to those who are suffering. Clothes and food drives can be led by children and can provide real relief to people who have lost their homes temporarily or permanently. Sending cards and letters of encouragement to children in a school that has experienced re-location can be helpful to children on both ends of the exchange. Even small actions can help children feel connected to positive responses that will help communities heal.

One caution I'll offer here is to make sure that efforts match what the suffering community needs. Sometimes disaster relief organizations and other volunteer efforts are overrun with goods that they cannot put to use. I recommend directly contacting an organization that is on the ground in the affected area and learning what is needed that children might meaningfully contribute to. If it is feasible for your school community, small fund-raising drives can also be incredibly helpful, especially for smaller, local organizations that may not have sufficient funding to do the work they are ready and able to do. Raising even $50–100 can contribute to good work, and children will see how combining their individual efforts/ contributions leads to a much bigger impact than each of them could have by themselves.

The NAACP has developed a toolkit to guide communities to adopt liberatory, justice-based practices at every point of response to the climate crisis, from prevention to preparedness to response and recovery from climate disasters. It is a detailed report, and well worth the time to read and understand as you work to integrate understanding and action around climate disasters in your classroom. Their framework is helpful as you evaluate efforts in your community and consider ways that you might involve your students in both understanding climate disasters in their community and engaging in hope-filled, collaborative action.[11]

In the coming years, no matter how quickly we act to reduce the harm of human-caused climate change, communities will continue to experience increasing and unpredictable environmental impacts. Schools cannot, by themselves, stop these from happening, but our role, as it has always been, is to help children feel safe, loved, knowledgeable, and able to act with hope as part of a supportive community.

Justice-Based Climate Science Unit Example: Caring for Our Community (Stream Clean Up) Aligned with NGSS for Kindergarten

(note that this is designed for children who have not been directly harmed by a natural disaster. When your school community experiences loss from a fire, storm, or other climate-related event, I urge you to draw on trauma-informed resources like those shared at the end of this section.)

Guiding Question: How can we help clean up a stream so it is healthy for plants, animals, and children and their families?		
Focal Disciplinary Core Idea Things that people do to live comfortably can affect the world around them. But they can make choices that reduce their impacts on the land, water, air, and other living things. (ESS3.C)	**Focal Science and Engineering Practices** Asking questions and defining problems Constructing explanations and designing solutions	**Focal Cross-Cutting Concept** Cause and effect
Engage with Concept and Community	• The starting point for this investigation will depend on the context in which it is situated. If there is a local issue that is causing children to be concerned about healthy water in their community, consider starting with a conversation in which children can share what they know, what they have heard, and what they are wondering. This will help them connect the classroom investigation to their own community and also begin to reduce their worry through knowledge-building in community.	

(Continued)

	• Once you are ready to begin the investigation, tell a version of the story that is used to open Chapter 9. Feel free to adjust the characters and the setting so that it will be meaningful for your students! This story works best if children and teacher are sitting together in a circle so that they can watch the story unfold in your model stream (a clear bin of water) as they follow along with the spoken story. As you talk about each new action that led to damage and pollution of the stream, add items to model that pollutant to the bin of water. • It can be helpful to have samples of the materials that you used to "pollute" the model stream so that children can touch them and consider what they might use to clean them up.
Define problem in need of action	• Explain to the children that they will work with classmates to form a stream clean up team. They will make a plan and try it with their own model stream. • Ask children to share ideas for how they might help to clean the polluted water. Consider using a think/pair/ share discussion model so that all children have an opportunity to generate, express, and listen to ideas. Sharing once while still in the whole group can help children get ready to share. Ideas with their stream cleaning team. • If at all possible, I recommend groups of 3 when doing this with young children, as I've found this maximizes opportunities for each child to play a role while also facilitating communication and collaboration. • Also go over safety and collaboration guidelines for this activity. Consider drawing symbols for some basic guidelines such as: make a plan before you start, listen to each person's ideas, try one idea at a time, and keep water in the bin (in a real stream you can't pour out the water!) *Note: because oil is the messiest material, the hardest to remove, and also often the most compelling, I usually do a separate, second hands-on investigation with more limited supplies for students to explore cleaning up oil from water. For this early phase, I use oil during the story telling but not in their initial materials exploration or in the bins that children will clean up.
Explore ideas grounded in place	• Show children some simple tools they could use to help them with the clean up. These might include some of the following: hinged chopsticks (the kind used by children who haven't yet learned to use separated ones) or tongs, spoons, a piece of paper towel, a small wire strainer, a paper coffee filter, a small piece of sponge, a magnet. Consider giving each team a tray with the materials they may use already on it. It's fine if not every team has exactly the same materials, since this is not a controlled experiment. Use what you have readily available.

(*Continued*)

(Continued)

	• Ask groups to draw a simple plan of what they want to try. Once they show it to the teacher, they will receive their model polluted stream and can try their ideas. A basic engineering design process that helps young children organize, develop, and evaluate their ideas is: PLAN: draw a plan of what you will try to clean the polluted water; list the plan in order TRY: try your steps in order and observe and draw what happened EVALUATE: worked well, how could you improve your method? This step works well as a gallery walk in which children can observe (but not touch) each group's first tries. REVISE: try again/ revise your plan to get better results
Design hope-filled actions	Once children have had an opportunity to explore and revise their ideas using a model stream, return to the story. Consider showing them a photo of a real stream that is visibly polluted. Ask students: how could they use what they learned from cleaning up a tiny model stream to help clean up real streams? Discuss what tools they might need. Students can then draw and share ideas for how they would work with other children and adults to clean up a stream.
Share and learn from community	How can your students be involved in clean water efforts in your community? They might be able to participate in a trash pick up to keep litter from entering a waterway. They might be able to help with planting that reduces erosion. They might make signs reminding people that rainwater carries litter and pollutants to the bay. Efforts that connect children's ideas from their in class experience to work in their own community helps to build the idea that they are knowledgeable and important contributors to their community's well-being.
Reflect and synthesize systems	Once children have engaged in this design challenge, continue to help them see how humans are part of the systems around us by asking them to consider the question of "where am I in the story?" Children can discuss and also draw their ideas about how they are connected to the systems around them and even those that seem far away from their lives. For instance, after hearing a read aloud about a rainforest far from their homes, they might consider the puzzle of how they are connected to rainforests. Coming back to the story of the stream is a good starting point for building these more complex connections. They might look on a map to follow a local stream all the way to an ocean that also touches the rainforest area, seeing how their local environment is physically connected to a far-away place. They can also think about how caring for their local community helps them understand the importance of the area around them, just as people and other living things in other places depend on healthy environments.

Recommended Children's Books

Calkhoven, L. (2020). *You should meet kids who are saving the planet* (M. Dong, illus.). Simon Spotlight.

Drummond, A. (2016). *Green city: How one community survived a tornado and rebuilt for a sustainable future*. Farrar, Straus and Giroux.

Marino, G. (2020). *We will live in this forest again*. Neal Porter Books.

Lindstrom, C. (2020). *We are water protectors* (M. Goade, illus.). Roaring Brook Press.

Villa, A. (2013). *Flood*. Capstone Young Readers.

Watson, R. (2014). *A place where hurricanes happen* (S. Stickland, illus.). Dragonfly Books.

Zissu, A. (2021). *Earth squad: 50 people who are saving the planet* (N. Le, illus.). Running Press Kids.

Notes

1 Haefele, A. (2022, March 16). *Environmental justice: The past and future in Oakland*. Ecoblock. https://ecoblock.berkeley.edu/blog/environmental-justice-the-past-and-future-in-oakland/

2 Romero, E.D. (2022, September 13). *"A lesson in discrimination": A toxic sea level rise crisis threatens West Oakland*. KQED. https://www.kqed.org/science/1980255/a-lesson-in-discrimination-a-toxic-sea-level-rise-crisis-threatens-west-oakland

3 East Bay Regional Parks District (n.d.). *Blue-green algae information*. https://www.ebparks.org/natural-resources/water-quality/blue-green-algae

4 Office of Human Services Emergency Preparedness and Response (2022, September 1). *Early childhood disaster-related resources for children and families*. https://www.acf.hhs.gov/ohsepr/disaster-human-services/children-and-disasters/children-and-families

5 Centers for Disease Control and Prevention (2022, July 18). *Returning to school after a disaster: Tips to help your students cope*. https://www.cdc.gov/childrenindisasters/index.html

6 Ready.gov. (2023, January 11). *Be a ready kid*. https://www.ready.gov/kids/be-ready-kids

7 National Association of School Psychologists. (2016). *Natural disasters: Brief facts and tips*. https://www.nasponline.org/resources-and-

publications/resources-and-podcasts/school-safety-and-crisis/
natural-disaster-resources/natural-disasters-brief-facts-and-tips

8 Kershner, J. (2022, February 28). *Restoration and managed retreat of pacifica state beach*. Climate Adaptation Knowledge Exchange. http://www.cakex.org/case-studies/restoration-and-managed-retreat-pacifica-state-beach

9 Buono, P. (2020, November 2). Quiet fire: Indigenous tribes in California and other parts of the U.S. have been rekindling the ancient art of controlled burning. The Nature Conservancy. https://www.nature.org/en-us/magazine/magazine-articles/indigenous-controlled-burns-california/

10 Farmworker Justice (2022, July). *Farmworkers and the climate crisis: Farmworker justice's environmental justice symposium summary report*. https://www.farmworkerjustice.org/wp-content/uploads/2022/07/EJ-Symposium-Report_07_14_2022.pdf

11 Steichen, L. (2021). *In the eye of the storm: A people's guide to transforming crisis & advancing equity in the disaster continuum*. NAACP Environmental and Climate Justice Program. https://naacp.org/resources/eye-storm-peoples-guide-transforming-crisis-advancing-equity-disaster-continuum

10

Toward a Pedagogy of Hope-Filled Action

"I have a question for the microtrash group. I'm wondering, how did it get to be microtrash? Like, did animals chew it up, or did people rip up the plastic and it just ended up getting blown to the bay?"

"Um, I think no to the animal part."

"But animals do eat the plastic. Remember that whale they found that had plastic all in its stomach? So maybe that's part of it."

"Yeah, true. But I think maybe it's just if the trash is there for a really long time it gets, like, smaller and smaller, but it doesn't really go away. It's just we don't see it as much because it's in tiny pieces. Like, if you see a whole plastic bag maybe someone will pick it up because they're like 'that's litter.' But if it's just the tiny pieces of plastic that we had to almost look at with a magnifying glass, and there all in the sand, who's going to pick that up?"

"But maybe if everybody did pick up the plastic bags then there wouldn't be any microtrash."

"My idea is, we need to tell people to just stop using plastic bags. And other stuff that turns to microtrash, like, well, I guess

DOI: 10.4324/9781003393535-10

other plastic stuff that people throw out. Oh yeah, like straws is what I was thinking. I think kids will listen to us. Cause, well, most every kid likes animals, plus they don't want to go to a gross beach with litter. And if we tell them the plastic is making there be climate change, I think that would work."

"I think, um, I want to build on what Lily said. I, well, I mostly agree because I agree kids will want to save the animals, so it will probably work. But if it's like a little kid and we tell them the climate is changing, they're going to be like 'what?' [the class laughs] Maybe we need to just tell them that plastic hurts the animals."

"Yeah, and later when they're older they can understand about fossil fuel and how that's what plastic really is, and how it never really breaks down and goes away."

The koosh ball we use to indicate the speaker and monitor airtime moved in rapid succession, as children built on each other's ideas and tried to make meaning of two months of work. The conversation above took place in my fourth-grade science class, at the end of our study guided by the question: is the San Francisco Bay healthy? Students had collected data from five different sites along the bay (with the help of parents coordinating of after-school trips, as we only had funding and time for one whole class field trip). They broke into expert groups to learn about environmental factors they had determined were important indicators of health, based on what we learned about the bay ecosystem and its connection to land-based communities. Groups studied water salinity, abundance of plankton and oysters, diversity and abundance of water plants and shell types, and presence of both macro and micro trash. They created expert group posters showing their data across the five sites, and we also created a whole class poster that compared the data across sites so that we could look for patterns.

Along the way, we learned about how the bay had changed over many years and also more recently. Students looked at maps showing how land had been created to expand bayside cities, impacting the size and features of the bay and the land that touches it. They learned about the critical importance of estuaries as the "nurseries of the ocean," providing sheltered, nutrient-rich areas for a wide variety of aquatic animals to reproduce. They explored how wetlands around the bay serve as a buffer between land and water, reducing the impact of storms on both land and water environments, helping keep water clean, and serving as habitat and feeding grounds for shorebirds and many other organisms.

At the end of the unit, students took up the question of what they should do to act on what they had learned. Their conversations, including the one above, led to action plans that ranged from visiting the kindergarten and first-grade classes to talk to them about being "plastic detectives" and always putting plastic they find in the trash to posters of wetland habitats that they thought would help other kids understand that these environments weren't "just mud." They explained to me that if kids all understood how important these places were, they would be able to convince adults to preserve and restore them. And if that didn't work, then the kids would grow up to be adults who would do the work. I believe them.

Justice, Community, and Science

I am an accidental science teacher. When I was completing the teacher credentialing process, I imagined being most passionate about teaching writing, since I had dreamed since childhood of being a novelist as well as an elementary educator. And I do truly love supporting children to develop their sense of power as writers and communicators. But in my early years of teaching, science is what pulled me in and what most reliably sparked students' curiosity and joy. I moved toward focusing on science first because of the sheer excitement that came from children exploring, experimenting, and creating. As I became more aware of how access to science education is deeply unequal, with children in high poverty schools and schools serving primarily children of color much less likely to have rich science experiences, I began to see science education as a social justice issue. But even within that framework, I hesitated to take up complex societal issues such as environmental justice and climate instability. My job as an elementary educator, above anything else, was to help children feel safe and supported in this world. I myself was terrified of what was happening to earth's systems, including the impacts on human communities. How could I teach children about this without them becoming frozen with fear and powerlessness when even I as an adult often found myself in that space?

My friend and colleague Gopal Dayaneni is a community organizer, climate activist, and educator. During a conversation in which another colleague and I were expressing climate despair, he reminded us that this framing treats disaster as the natural and inevitable progression of our current state, but that communities engaged in collaborative care and action are able to and are transforming the world. With his permission, I share part of his recounted response here:

> While there are many aspects of climate disruption that we cannot stop from happening because they were set in motion a long time ago—we absolutely can change how we experience them—and through relationships of care, cooperation, consent and compassion—we transform the experience from catastrophic suffering to shared struggle. As educators, it is our responsibility to not give our students simple solutions that make them feel good about themselves for fear of despair or to just *inspire* them into action. It is our responsibility to accompany them on a journey of collective agency and transformation. To remind them that hard and bad are not the same thing and that there is nothing inevitable about how the world works today or how the crisis will unfold. To remind them that it is through our cultures, ancestries, and experiences that we will harbor the knowledge to remember our way forward to a better, more just and resilient future. Everything we need to navigate the crisis well is here, now. It must be uplifted, defended, nurtured and expanded.[1]

It is this perspective that allows us to act not from fear but from love, and not alone but in community with our students, their families, educator colleagues, and our larger community When we engage with children in hope-filled, justice-based learning about climate science and the global systems it entails, they experience the power of collective understanding and action and are able to dream of and bring to life responses that come not from despair but from the belief that their ideas and actions matter and have the power to change their community and our world for the better.

If you are just beginning the journey of infusing ideas of climate science and climate justice into your elementary classroom, the process may seem daunting even if you believe that children are both capable and eager to take up complex scientific and social ideas. If you teach in a setting with a scripted science curriculum that does not focus on systems, place-based learning, and hope-filled action, it may be hard to envision how to create space for a different approach. If your school de-emphasizes science teaching and/ or requires that most of the day be spent on literacy and math, it is similarly challenging to create time for more integrated learning. And while I've tried throughout this book to provide an introduction to key ideas and issues in climate science and climate justice, many of us who work in elementary classrooms worry that we do not know enough about the science to teach it well, and we worry about providing wrong or misleading information.

These are all very real, structural issues in our education system, and I won't be able to solve them in the final chapter of this book. But just as it is in the climate justice space, community and collective action among educators and the communities they serve can build our hope, stamina, and capacity for enacting and sustaining change. So I will end this book with a few ideas about how to build our individual and collective capacity to create the learning conditions needed for children to grow into engaged, empowered community members who address the challenges of living as part of our transformed planet.

Start with What You Know and Love

One of my favorite (and most ambitious) science units in the early years of my teaching was building soda bottle ecosystems. These multi-level model habitats allowed children to explore the interaction of living and non-living things in both a water and land-based environment. The unit was designed with a focus on systems, so when I decided to more explicitly engage children in issues of climate justice, this seemed like a promising starting point. Instead of starting from scratch with a unit I had never taught, I used this beloved unit that I had taught at least ten times to ground me as I took pedagogical risks.

First, I worked to make the unit more explicitly place-based. The exploration was written in a way that could be enacted anywhere, as most published curricula are, but the readings about organisms and environmental issues were focused on the Chesapeake Bay. When I first started teaching, my school community was fairly close to this area, but by the time I decided to move toward more climate-infused teaching, I was in a large city in California. My school was in a densely populated, urban neighborhood, but we were within walking distance of a small creek, a park, and part of the San Francisco Bay. So we began the unit with a walking field trip, looking for evidence of plants and animals and how they survived where we live. As we built the model ecosystems, we learned about local species and people who were working to preserve, clean, and expand natural habitat.

I also greatly increased the time spent and ways in which we explored energy and matter transfer between different parts of the system. This was the scariest part for me because, while this was an area of scientific confidence for me, I was worried that the content would be too challenging and children would lose some of the joy they experienced as they watched plants sprout, crickets hide in the leaf litter, snails give birth in these tiny ecosystems. This is where I learned the power of story, focusing on narratives rather than lists of facts to help children understand how matter is constantly flowing within and among the plants, animals, and non-living things.

It was another couple of years before I decided to add a study of the carbon cycle as a supplement to this unit, a process I discuss in Chapter 3. And it took a few tries to get those lessons to the point where they were a good combination of scientifically accurate, engaging, and tied to hope-filled action.

I had been a science teacher for about ten years when I began making these changes, and so this example is complex. My newer first-grade colleague, who I mentioned in Chapter 2, started with a classic early childhood experience, raising caterpillars, something she had done with children as a student teacher and was eager to bring to her classroom. We worked together to embed this into a systems story, exploring local plants and pollinating insects, and helping children consider their role in this system. A colleague who loved teaching her students about weather added an engineering design problem similar to the one that I discussed in Chapter 6, to help her children explore how weather and humans are connected and to engage them in innovating to design more storm-resistant shelters.

Often it makes the most sense to layer exploration of local community and systems into your required science curriculum. You might start small by simply adding local stories and place-based exploration to place-agnostic curricular materials. As you observe how students respond to and take up these stories, you are likely to feel more comfortable veering from the script, adjusting investigations to be more explicitly linked to your community, the shared knowledge there, and the challenges and points of hope tied to your children's lives and experiences.

Find the Stories and Help Children Ask: Where Am I in This Story?

As we've seen in several of the examples of practice throughout this book, children are eager to know the stories of their place and also those of other communities. Story is a powerful means of sharing challenging concepts in ways that help learners make connections between things they already know and new ideas. More broadly, story is how we make meaning, both in our minds and in community with others. The stories that my family has shared with me throughout my life help to shape my knowledge, my values, and how I see myself as part of a community over time. This is true in families but also in other social networks including the classroom community.

Learning the climate stories of your students, their families, and your community is a powerful way to bring climate justice into your teaching practice. If your class is studying the water cycle, consider asking children to interview their parents to learn stories of when they might have struggled with too much or too little water. The stories from children living in a farming community will be quite different from those who have experienced urban flooding. And there are likely parents and elders in some of your children's families who grew up in very different environments and can share stories of humans and water that expand our shared knowledge beyond the local community. I did this activity many years ago when I was teaching third grade in Atlanta, Georgia, and one child came back with a story of her mother spending summers at her grandparent's home, a sharecropper's cabin only about 100 miles away from the city, but even late in the 20th century the bathroom was an outhouse and the only water access was from a well that was not always safe to drink from. Children looked on a map to find the tiny town she was talking about, and this led to a discussion of clean water as something all people need access to. But the mother's story was one of fond remembrance of visiting her grandparents, not one focused on deprivation, so learning this story helped the child and her classmates consider how people innovate and find solutions to challenging situations.

Access to clean water is an environmental justice issue that is being worsened by climate change, as increased droughts lessen availability and stronger storms lead to increased runoff and pollution of potable water sources. The elders who had worked as sharecroppers in my students' family faced immense inequality in access to resources due to racism. So I want to be clear that I am not proposing we use stories to romanticize the inequality and environmental racism. However, the fact that my student's mother centered the joy of being a child, beloved by her grandparents and thus remember fondly the time spent in their home, clearly filled her daughter with pride and connection to her elders. And passing down the stories, experience, and love of those who came before us helps create a web of hope that shows us we are not alone or the first to engage in struggle.

We can also use story to amplify the people and perspectives that are often made invisible in official school texts, including indigenous communities. Who are the indigenous people whose land now houses your school? What are their stories, past and present, of how they interact with and live in relationship with the local environment? What indigenous-led efforts are happening in your area that can help children come to understand their full, interconnected community? Because indigenous people have been driven from their lands, killed, and forced to assimilate throughout US history, stories of indigenous people in this country often reside in the past tense. But of course that is not accurate, as there are vibrant indigenous communities throughout the country, and their knowledge and ways of understanding interconnected earth systems is vital to taking up the challenges caused by anthropogenic climate change. Indigenous communities are often particularly vulnerable to the impacts of climate change, due to their forced settlement on environmentally precarious lands and high poverty due to limited economic opportunities. But indigenous communities also use generations of knowledge about the land and human's relationship to it to engage in sustainable approaches to farming, water use, fire management, and community-based responses to climate crises. Seeking out and learning alongside indigenous elders in your community creates a more inclusive story of what it means to engage in climate justice.

As I discussed in Chapter 2, stories are powerful by themselves, but helping children connect themselves to stories that may at first feel outside of their own experience makes them even more effective teaching tools. I love to ask children, "where are you in the story?" Their answers often range from the literal (I'm like one of the kids who was scared of the fire) to the aspirational (I'm one of the people helping with the burn) to the poetic (I'm one of the new plants that grows up after the fire). Sometimes this part of class discussions can get a little silly, but more often, children considering how they fit into the story helps them connect to the narrative in new and sometimes transformative ways.

Find Your People

Bringing developmentally sensitive climate change and climate justice education into our classrooms is one of the most important things we as educators can do in this time of immense planetary change. Yes, we need to prepare our next generation of leaders, but we also need to help communities feel connected to and empowered to enact hope-filled change right now. Children cannot enact government policy, but they can be part of a broader movement to center planetary health in all that we do. And we as educators play a critical role in nurturing this perspective.

Sometimes, though, the work is exhausting and lonely. If you teach in a school where the science curriculum feels disconnected from place and not at all focused on systems, the idea of overhauling it to better support climate-aware learning is daunting. If science and social studies have been pushed out of the curriculum altogether, or sidelined into a once a week "extra," figuring out how to integrate climate education into your educational program may feel nearly impossible. And if you teach in a community where some folks are hostile to the very notion of acknowledging and addressing anthropogenic climate change and the justice issues it has caused, choosing to engage in climate education can feel risky.

There is no quick fix for any of these issues, but all of them are easier to take up from a position of hope and power when working as part of a community. This might look like working with other like-minded teachers at your school to identify, develop, and share resources. Some of my most rewarding teaching experiences have involved teaming up with teachers in different grade levels or with different subject expertise to develop a shared project, so that we are able to share both ideas and workload. I also encourage you to find the people in your area who are already engaged in climate education work. There are people in nearly every community working in some way on issues of climate education, but we often aren't well connected to each other. Finding an organization or group doing work that you care about can provide you with thought partners, resources, and community. I have worked for several years with scientists and educators at the San Francisco State University Estuarine and Ocean Sciences Center[2] and the SF Bay National Estuarine Research Reserve,[3] both associated with the university where I teach, but whose existence I barely knew about for many years. Once we connected, one project led to another. They have led field trips for both my elementary students and my teachers in training, we have developed curriculum units together, and they keep me connected to current research and other community partners. Their research and field education inspires and energizes me, and my classroom-based work expands their perspective and reach. More recently, I've been privileged to work with like-minded colleagues on my campus around issues of climate change and climate justice education spanning from early childhood through graduate study.[4] Collaborating on this challenging work keeps us all more hopeful than downtrodden, as we lean on and encourage each other and remind one another of the importance of this work.

What organizations are already working to advance climate action in your community? I encourage you to connect with regional park systems, museums, local colleges, and associations of science and environmental educators. Finding or developing a group of people who help you remember why this work matters and who are committed to working in community, is key to sustaining our own work through collective action as well as working toward systemic change.

We Are Teaching Not Only for Today but for the Next Thousand Years

Teaching is a sometimes messy blend of the mundane and the divine. Daily practice with phonemic awareness matters, and so does helping children practice empathy and love. Systems to keep the classroom running smoothly and safely matter, and so too does being open to the surprising, brilliant ideas and questions that children offer and can change the course of the school day or even the whole year. Helping first graders to be ready for second grade matters, and so too does developing the skills, stances, and knowledge that will allow them to be engaged community members for a lifetime.

The work we do in our classrooms matters today, and it also matters for the future. Children who experience being part of a hope-filled community committed to acting for climate justice will carry this into their later lives. The experiences they have in elementary school can inspire them toward future study and action toward addressing climate change in systemic ways. We know that the elementary years are critical for sparking and growing children's interest and feelings of competency and belonging in STEM fields, and we also have the privilege of developing children's ideas about what it means to "do science."[5] By approaching climate change education through a lens of understanding the earth as interrelated systems, grounding learning in place-based learning, and engaging in hope-filled action, we help children understand that knowledge grows in collaboration and community with others, and science and social action are inextricable.

There is so much in our society that needs to transform to ensure resilient, sustaining communities for all living things on our planet. What we do today and tomorrow with a small group of children in our classroom can feel tiny and insignificant in the face of these challenges. And yet building communities of knowledge-building and hope-filled action that stretch beyond our classroom, connecting our work with those in our community who are also engaged in justice-based learning, advocacy, and action, is how transformative change happens. I hope that you will use some of the ideas from this book as you continue to build hope- and justice-filled community in your classroom, school, and the world beyond, and know that what you do matters tremendously for your students, for our society, and for the future of our planet.

Notes

1 Learn more about Gopal Dayaneni's work here: www.move-mentgeneration.org
2 San Francisco State University Estuarine and Ocean Sciences Center: https://eoscenter.sfsu.edu/
3 SF Bay National Estuarine Research Reserve: https://sfbaynerr.sfsu.edu/
4 https://climatehq.sfsu.edu/climatehq/climate-justice-leaders-initiative
5 *Early Exposure to STEM and Its Impact on the Future of Work.* Purdue University. https://gems.education.purdue.edu/wp-content/uploads/2019/02/STEM_in_Schools_v1-2.pdf